Also by Robert Vaughan

The Crocketts

Western Saga, Volume 1
Western Saga, Volume 2
Escape From the Devil (Book 9)
Justice Of The Gun (Book 10)

Western Adventures of Cade McCall

Long Road To Abilene
Cade's Revenge
Cade's Redemption
Cade at The Wall
Army Scout

The Faraday Series

The Iron Horse
The Gold Train
The Trackwalker
Train of Glory
Buffalo Train

Random Thoughts Of An Old Writer: A Memoir

Robert Vaughan

Paperback Edition
Copyright © 2021 Robert Vaughan

All rights reserved. No part of this book may be reproduced by any means without the prior written consent of the publisher, other than brief quotes for reviews.

This book is a work of fiction. Any references to historical events, real people or real places are used fictitiously. Other names, characters, places and events are products of the author's imagination, and any resemblance to actual events, places or persons, living or dead, is entirely coincidental.

Published in the United States by Wolfpack Publishing, Las Vegas

Wolfpack Publishing
5130 S. Fort Apache Road 215-380
Las Vegas, NV 89148

wolfpackpublishing.com

Paperback ISBN 978-1-64734-565-5
eBook ISBN 978-1-64734-552-5

Random Thoughts
Of An Old Writer

Prologue

Over the course of my long writing career, several people have suggested I should write my autobiography. All right, it was my wife and one Facebook friend. That's several, isn't it? Actually, my wife didn't exactly suggest I do so, but she finally came around when I said I might.

This isn't an ordinary autobiography, however. I don't start at the beginning of my life, when my mother went into labor and I was almost born in a chicken coop. And I don't start at the very, very beginning, god forbid, when I was the single one out of 100 million or so sperm—yes, 100 million; you can Google it—that won the great egg-hunt race, though that experience has taught me that if you can beat those odds, you can do just about anything in life.

No, this isn't an A-to-B narrative. Rather it is slices of my life—vignettes placed in vaguely chronological order. That is why I am calling it Random Thoughts of an Old Writer. It isn't arranged by chapters but by titles of the vignettes I am sharing with you.

I have further broken it down into four sections of my life:

Youthful Adventures, Army: Enlisted Man, Army: Warrant Officer, and A Civilian at Last.

I hope you like it and tell all your friends. But if you don't, just do me a great favor and keep it to yourself.

Youthful Adventures

Pilot Error, or Mechanical Failure

I am the oldest of three sons. My brother Tommy was in the middle, and Phil was the youngest. Let me first share something about my middle brother.

Tom was a rather unique individual who was afraid of nothing. Here's an early example. When I was about 7 and Tommy was five, I built an "airplane" from a wagon, with a wing made from an ironing board that I tied to it. I was absolutely convinced it would fly, and Tommy kept begging me to let him fly it first. But I built it, and the first flight would be mine. We took the contraption up onto the roof of the garage. Once we got there and I looked down, I decided to grant Tommy his wish. Seated in the contraption, he raced down the roof with a broad smile, then fell to the ground. Thankfully, he wasn't hurt, and surprisingly, he wasn't angry with me.

Saving the Ants

Phil is my youngest brother and the only one of the three of us to have been born in a hospital. When I went there to see him for the first time, Mother held him up and introduced him. "This is Phillip. What do you think of your new brother?"

"He's fine," I said. "But where's Phyllis?" Mother had already given birth to two sons, and she was so certain that the third child would be a girl that Tommy and I fully believed that you could just order the sex of a child, and that Mother had ordered a girl.

Dad was drafted shortly after Phil was born, and we followed him to Fort (then Camp) Rucker, then to Fort Sill. We made those long trips in a 1941 Ford, with Tommy and me sleeping on a pallet in the back seat and Phil sleeping on a pillow in the seat next to Mother. There were neither seat belts nor car seats in those days.

On TV now, there are all these touching pictures of fathers returning from the war and reuniting with their children. My first view of Dad, when he came back, was when I got up in the middle of the night to go to the bathroom and saw him standing at the toilet stool. That probably would not have made one of those heart-warming vignettes they like to show on TV.

At least I knew who he was, and I welcomed him home. Not so Phil, who cried every time he saw this strange man in the house. To get him to eat, Mother would have to turn his high chair around so he didn't have to look at Dad.

Phil has always had a sense of compassion for other life forms. Being kind to animals is one thing, but Phil took it a step further. When he was very young, I saw him sitting out on the sidewalk. I thought nothing of it, but several minutes later, he was still there.

Curious what he was up to, I walked over and saw he was holding a stick against the concrete walk.

"Phil, what are you doing?"

"I'm turning the ants so they won't go out in the road and get run over." Now, this from someone who had two older brothers who loved playing "war" against the ants, smashing as many as we could with our finger.

Because Phil was five years younger, I didn't have as much interaction with him as with my other brother. Tommy wound up in the DEA, I did three years of combat in Vietnam, but in many ways, you could make a case for Phil having the most courage of the three of us.

Let me tell you why.

Phil was in the third grade and playing with the other kids as they were waiting for the bus to take them home. Someone jumped on his back and, when he fell, he broke his arm. I had broken my arm at about the same age, so initially I didn't think that much about it. But there was a tremendous difference between the simple fracture I had, and the life-changing fracture Phil suffered. In the almost 70 years since that happened, he has not been able to move his arm at the elbow.

I remember lying in bed at night, listening to Mother

and Dad talking.

"I just don't know what that boy is going to do," Dad would say. "With his arm like that, he won't be able to do anything. How is that boy ever going to be able to make a living?"

Ha! Phil did not let a little thing like a frozen elbow stop him. He is a handyman's handyman and can fix just about anything. He became an AHC, (Architectural Hardware Consultant), which in the construction business is a highly respected title, and he made good use of that title, first as a well-paid employee, and eventually to build his own million-dollar business.

Dad had always equated making a living with being able to handle d-pound bags of sugar so you could load and unload trucks. As it turns out, using your mind is better than using your muscles.

Although Phil is 77 years old, to me, he is still my little brother, and I'm proud of him.

The Promise

Our hometown of Sikeston, Missouri, was flanked by a series of drainage canals that had been dug in the first decade of the 20th century to drain the swampland and produce some of the richest farmland in the country. We called them "ditches," and as kids we would fish in them. But we weren't supposed to swim in them, as polio was rampant then, and swimming in such unsanitary water was considered a leading cause.

But Tommy and I did. Shortly after he and I went swimming in "First Ditch," Tommy got very ill. I just knew it was polio. Mother asked if we had gone swimming in the ditch, and I lied and said no.

"Well, I'm going to take Tommy to see Dr. Urban. I'm worried about him."

As Mother was getting ready, I went into the bedroom to see my brother.

"Tommy, Mother is going to take you to the doctor to see if you have polio. You probably do, and you're probably going to die, but please don't tell her we went swimming in the ditch. You know she won't punish you, because you're going to die, but she would punish me, so please don't tell her."

"I won't," Tommy promised in a pitifully weak and self-sacrificing voice.

It turns out that all Tommy had was a cold . . . but that promise cost me for the rest of the summer. Anytime there was a choice to be made on anything, he got to choose first.

Many, many years later, when Tommy came to Fort Eustis, Virginia, to visit me, we started discussing where to go out for dinner. Tommy reminded me that he had "first choice."

"What do you mean, you have first choice?"

"Because when I was dying with polio, I didn't tell Mother that you made me go swimming in the ditch," he said triumphantly.

In high school Tommy was a good athlete and also had the dark hair and brown eyes that so many of the girls found attractive. Often one of the girls would call me and ask if Tommy had a date for Saturday night.

"Yes, he does, but I don't."

"No, I don't want to go with you, Dickey, I want to go with Tommy. Would you tell him I called?"

I never did.

The Kiss

My next-door neighbor, Neil, was a fifth-grader, two years older than me. Neil let me know in many ways, subtle and overt, that he was older, delighting in telling me such things as: "You can't be a member of the schoolboy patrol until you're in the fifth grade, because you have to be very . . ." he paused before saying the word he had just learned, "responsible. I mean, what if you said some first-grader could cross the street . . . but a car didn't stop? Why, you would have to be very brave and run out into the street and pull him back to save his life."

Neil ran his thumb over the badge of the white Sam Browne belt he was wearing. "You have to be brave to be a schoolboy patrol," he said.

I was very envious. At Sikeston's South Grade School, those in the schoolboy patrol were akin to the football players in high school; everyone looked up to them. They were well aware of that, too, and you would see them at recess, gathered in little clusters, easily identified by the white belts and badges. They shared "inside" stories such as: "You have to really keep an eye on Gerald, because he'll step out in the street if you don't watch him."

"Bobby is just as bad," another would answer. "That's why I hold my arms straight out, just to keep someone from running out into the street. They're so young—you have to watch them like a hawk."

It wasn't just being a member of the schoolboy patrol that made me envious of Neil. As a fifth-grader, he was much more worldly than I was. Of course, I wasn't aware of the term worldly, but I was certainly aware of the concept.

"Dickey, have you ever kissed a girl?" Neil asked one day.

"Sure, I've kissed a girl."

"Who have you kissed?"

"I've kissed my mother."

Neil laughed. "She isn't a girl, she's your mother."

I tried to respond with my grandmother and a couple of aunts, but Neil said that didn't count either.

"I mean a girl, a young girl your own age, and someone who isn't kin to you. I've kissed Melba. She's in the fifth grade. Have you ever kissed a girl?"

"No," I had to answer, crushed by the fact that, once more, Neil had gotten the better of me.

The next day in school, I was sitting at my desk thinking about what Neil and I had been talking about. He was right. If indeed my mother, grandmother, or any of my aunts didn't count, I had never kissed a girl. How could I ever hope to consider myself equal to Neil? Why, he could even cross Kingshighway on his bicycle, something my mother wouldn't let me do.

Suddenly, the voice of my teacher, Miss Stewart, interrupted my musings. "Now, boys and girls, I want you to turn to page 15 in your Third Grade Readers and read the story about Billy and The Magic Horse," she said. "After you have had time to read it, we will talk about it."

I had already read the story. In fact, I had already read every story in the entire book, so the next 15 minutes was going to be very boring.

As I sat there pretending to read, I looked just across the aisle and saw Patty studying the story. She was seated in such a way that her elbow was at the edge of her desk . . .

Sticking out into the aisle . . .

About 18 inches from me . . .

Just hanging out there . . .

Patty was a girl, I told myself. Gathering my courage and looking around to make sure none of the other kids saw me, I leaned across the aisle and kissed Patty on the elbow.

"I saw that, Dickey Vaughan!" Miss Stewart said in a booming voice. "You are going to be kept in for recess, young man."

I looked over at Patty, and she was rubbing the spot on her elbow. She was also smiling at me.

I regretted having to stay in for recess, because we were in the middle of a weeklong softball game with Miss Caine's third-grade class, and I was sure to come to bat today. But . . . I had kissed a girl!

I could hardly wait to tell Neil about it.

The 100-Year-Old Turtle

When I was in the fourth grade, Mrs. Margrabe read our class a story about turtles. She said that turtles could live for a long time, and that some lived to be 100.

That weekend, Tommy Murchison and I went out to hunt for turtles. They were easy to find, so it didn't take us very long.

"How old do you think he is?" Tommy asked, holding the one we had located.

"I don't know. Mrs. Margrabe said they could be a hundred years old."

"I wonder, was he here during the Civil War?" Tommy asked.

As we continued to speculate, we examined the turtle until we found what we were absolutely certain was a scar that had to be from a mine ball, fired at it by some Civil War soldier, though whether he had been attacked by Yankee or Rebel, we weren't sure.

I'll be honest with you, I don't know which one of us got the idea, but it was received with equal enthusiasm by both parties. We decided to "declare" this turtle to be 100 years old, and we did that by a very simple method. We carved into the turtle's shell the year 1846–a hundred years prior to our little experiment—then we let him go.

Over the last 74 years, I have wondered, from time to time, if that turtle is still with us. And if so, has anyone ever found it and marveled at how old it was . . . as was clearly (if fraudulently) indicated by the carving on its shell?

I hope so. And I truly hope that someone, somewhere, has told the story, hundreds of times over the past 74 years, about how they found a turtle that was over 100 years old.

Memories

One good thing about being a writer is that I am able to pull up memories from the distant past and relive them. There can be a downside, of course, if the memories are particularly unpleasant. But if the memories are pleasant, recalling them can be almost as enjoyable as when I experienced them firsthand.

This particular memory goes back well over 70 years. My mother's family was from Jackson, Mississippi, and we used to go there to visit my grandparents and my mother's many sisters and brothers. Often we would take the train, which would leave from the Frisco Depot in Sikeston at three o'clock in the morning. Standing out on the brick platform, I would see the light of the train approaching from the north, even before it crossed over West Malone Street. They were steam engines then, and as the big locomotive drew near, it would be gushing smoke and spewing steam from the drive cylinders. The big wheels, driven by the thrusting piston rods, would rumble by, so heavy that it would make my stomach shake. Burning cinders would spill out from the firebox, creating a long, glowing path between the tracks.

Finally, almost to my surprise, the train would come to a squeaking, hissing stop. Though motionless, it was far from quiet, as water would be burbling in the boiler and the overheated journals and bearings would pop and snap. I would gaze through the lighted windows at the passengers as they slept, read, or stared with bored expressions at the small depot. I always felt a rush of excitement when the conductor stepped down, glanced at his pocket watch with a sense of importance, then shouted, "All aboard!"

Getting to Jackson was exciting, but the real fun would start once we were there. Uncle Ernest always managed to take at least one day off to show a good time to we kids: me and my brothers Tommy and Phil, and my cousin Ronnie, who was halfway between Tommy and me in age.

One visit in particular stands out. We began the day by going to visit the zoo at Livingston Park. I loved the monkey island, while Tommy and Ronnie liked the snakes. After the zoo, we had lunch—I ordered my usual, a Coney Island—then we went to a movie. On this day we saw Abbott and Costello in *Keep 'Em Flying*. That was a double treat for me: Abbott and Costello were always funny, and I loved anything to do with flying. Following the movie we went to an ice-cream parlor, where I had my very first banana split.

Then we were off to the Tower Building, as the Standard Life Building was called then. At the time this 22-story building was not only the tallest in Jackson, it was tallest structure I had ever seen, much taller that Tom Dumey's silo back in Sikeston. On the observation deck there was a brick bannister about shoulder high and about three feet wide. To my shock and horror, Ronnie climbed up on it and walked from one end to the other! I was so frightened that I nearly threw up, and to this day, I am greatly bothered by heights—other than in an airplane. I blame Ronnie for that.

Our final stop of the day was at the old capitol building and its museum, which to my amazement had a mummy on display. The mummy had been donated to the state in the 1920s. But a thorough examination in 1967 proved it to be a fake. That was well after my visit, and when we saw it in the late 1940s, it was still a mummy! So as far as I'm concerned, I have seen one.

Stan Musial

I am an unabashed St. Louis Cardinal fan. I can still name the starting lineup of the Cards from 1946 to 1950. I didn't miss one Cardinal game during that time, listening to Harry Carey's live radio broadcast. It was easy to follow them and still play outside, because there were no air-conditioners in those days. Every house had the doors and windows open, and they all had the radio on, tuned to the Cardinal games.

All of us had favorite players. My brother's favorite was Enos Slaughter, Kenny Kern's was Whitey Kurowski, David Lewis's was Marty Marion—but I suspect that was because David's middle name was Marion. My favorite player was Stan Musial. During the 1949 season, Musial was locked in a head-to-head contest with Ralph Kiner of the Pittsburgh Pirates for the most home runs. I religiously kept up with them: Musial would go ahead, then Kiner, then Musial. Once while playing kick-the-can, I was hiding in a bush behind the house of Mr. McVey, who of course was listening to the game. Musial hit a home run, which not only put the Cards ahead of the Brooklyn Dodgers, it put Musial ahead of Kiner. I couldn't keep quiet, and shouted, "Musial hit a homer!" That gave away my hiding place, and Ray Kelly beat me to the can.

During the wrap up of the Cardinal half of one of the innings, Harry Carey said, "Kurowski died at third."

Tears began flowing down Kenny Kern's face at hearing the fate of his favorite third baseman, and he said in bewilderment, "He died?"

"What?"

"He died! Kurowski died!" Now the weeping was audible.

"Kenny, What are you talking about?"

"Harry Carey just said that Kurowski died at third base."

"That doesn't mean he's dead, Kenny, just that he was on third base when the inning ended."

"Of course he was on third," Kenny informed me as if I didn't understand. "He plays third base."

"Think about it, Kenny. The Cards were at bat. He wouldn't have been playing third base if the Cardinals were at bat. It means he was left stranded as a runner."

"Oh, yeah, you're right."

Now my story jumps more than 50 years, when it had been arranged for me to have dinner with Stan Musial to discuss the possibility of ghosting his autobiography. I found myself remembering those early years and the first time I ever actually saw the Cardinals play. It was in Sportsman's Park, where I saw on the field all my heroes, their white and red uniforms shining against the green grass. That single instant, when I first looked down on the field, is one of those magic moments that will be locked forever in my memory. And all those years later, when Musial stuck out his hand to greet me, he thought he was shaking hands with a 60-year-old writer. He had no idea he was shaking hands with an awestruck, 10-year-old kid.

We met in his hotel, and he invited me to a meeting room to talk, but when we stepped inside, several of the maids were there having lunch. Embarrassed, they jumped up and started gathering up their dishes.

"No, no, no . . . stay here, please," Musial said. "We'll go somewhere else."

We ate in the dining room, where Musial had several occasions to interact with other staff members, and in every interaction he was the most gracious man I believe I ever met. The meeting was wonderful, filled with many great vignettes, including the fact that one of his home runs had not been counted in 1949. It had occurred in a game that was called for rain and never completed because it had no effect on the final standings. Had it been counted, he would have shared the home-run title that year with Ralph Kiner. Instead he had to be content with the most singles, doubles, and triples, and the highest batting average.

At the end of the dinner he gave me an autographed baseball. I said with a smile, "Why don't you write on it that this is the ball you got your 3,630th hit with?"

He smiled. "I couldn't do that. I have that ball."

I wish I could say I ghosted his autobiography, but it was his inherent decency that got in the way. "There's been enough written about me," he said. "For me to write my own story, even if you wrote it, would be sort of like tooting my own horn. And I really just wouldn't feel comfortable doing that."

Old Friends

When I was five years old. I lived across the street from Gerald and Ray Kelly, the earliest friends I can remember. Gerald is gone now, but I'm happy to say that Ray is still around. I also lived across the street from Mr. Clark, who was a Civil War veteran. I have to admit that being a Civil War veteran meant nothing to me then. I knew him only as a very old man who used a cane to walk to the corner and back. I also confess that I don't know which side he was on, as this was Missouri, a state where brother literally fought against brother.

I remember World War II vividly. I remember the air-raid drills and blackouts . . . and being absolutely certain that Japanese airplanes would come from the West and German airplanes from the East, to rendezvous over the small town of Sikeston because, in my young mind, we had to be a prime target.

I remember the excitement of the new cars coming out after the end of the war, and I was very disappointed that we didn't get one until 1949—a shiny new Plymouth. In

1950, Little League baseball came to Sikeston, and I played catcher for the Dons. I didn't like the team name, which meant nothing to me (though I later learned it meant our team wasn't sponsored but was the result of donations). The good part about playing for the Dons was that our coach was Bill Puckett, who was "Mr. Baseball" in Sikeston.

It was about then that I saw a television for the first time, not installed in someone's house but in the back of a radio repair shop. I stood staring at the blank screen, totally enthralled with the concept of seeing moving pictures in our own living room. A neighbor had the first working TV I ever saw, and often on nights when he called to say "the picture is clear," we would join a dozen or so others, sitting in chairs that were arranged like seats in a movie theater, watching whatever was on.

In 1953, when I was 15, I joined the National Guard. I went to summer camp three times (two weeks of active duty) the first time at Camp Ripley, Minnesota, and the next two times at Camp McCoy, Wisconsin. We went by troop train, and once when the train was stopped on the track in St. Louis, someone shouted, "Whoa! Fellas! There's a color TV!" We ran to the windows of the car and saw, through the open window of an apartment building, a color television. We watched in wonder until the train got under way again.

While seeing that newfangled electronic wonder was certainly memorable, it wasn't the greatest memory of my time in the National Guard. One of the duties of the Guard was to provide military funerals, which at the time were mostly for World War I vets. I was often on the squad that fired the salute over the grave . . . and a few times I had the honor of playing *Taps*.

Beginning of my Writing Career

I am often asked how I got into writing. I suppose I could say that my writing career began in 1948. I was in the sixth grade and in the principal's office when I first saw a mimeograph machine. One of the school secretaries was turning the crank, and page after printed page was flying off the drum.

I was so mesmerized that I stood there watching it in such awe and didn't realize Mr. Travelsted was talking to me.

"What do you need, Dickey?" he asked several times before I responded. I gave him the note Mrs. Willum had given me, then I started questioning the secretary about the mimeograph. She showed me the stencil and told me how it worked, then gave me a stencil.

I knew then that I was going to be published. I p my best friend, Tommy Murchison, as my partner in the enterprise. Tommy could print beautifully and could also draw very well. I was lacking in both of those skills, but I could write the stories. And thus the publishing firm of *Willum's Gazette*, Vaughan and Murchison Publishers, was born.

After I wrote the stories, Tommy printed them very carefully on the stencil. Our sports column reported the scores of the softball games played at recess. I wrote about the adventures of the schoolboy patrol, or I would pick a book from the library and review it. The school secretary printed out about 200 copies each week, and we provided every room in South Grade School with enough issues to satisfy the demand.

It was Tommy who came up with the idea of holding a poetry contest. The winner would have their poem published in *The Gazette* and receive two tickets to the Malone Theater. At the time the cost of admission was 10 cents for anyone 12 or younger, which meant that we would have to come up with 20 cents.

We had at least 30 submissions, and the poem by Sarah Potashnick was very good. She wrote of the coming of spring, the freshness of the air, the beauty of the flowers. The rhyming and meter was excellent.

"Here's our winner," I said, showing the poem to Tommy.

"Huh, uh, this is the winner," he insisted, holding forth a different one.

Taking it from him, I read the poem aloud: "Old Mother Hubbard / went to the cupboard / to get her dog a bone. / But there wasn't anything there, / so the dog went home."

I looked up at him incredulously. "What do you mean this is the winner? This poem is awful! Sarah is the winner."

"Look who wrote this one." Tommy pointed at the author's name: Vernon M. "Think about it, Dickey," he said. "Vernon can beat us up. Sarah can't."

Vernon got his poem published and two theater tickets.

Sarah, if you are reading this, I know it is 70-plus years too late, but you actually won the contest. However, it is not possible to award you two tickets to the Malone Theater, as it is no longer there. But if you send me your address, I'll mail you the 20 cents.

Our First Television Set

I was in the sixth grade when I heard someone say that there was a television set at the local radio shop. I had heard of television, but I had never seen one, and I could hardly wait until school was out. Barrett's Radio was on Greer Street, just across from the shoe factory, and as I lived in the 800 block of Greer, I would be riding my bike right by the shop on my way home.

I leaned my bike against the front of the store, went inside, and asked, "Is it true? Do you have a TV set here?"

"It's in the back, but it isn't hooked up."

"May I go look at it?"

"Sure, but it isn't hooked up," he repeated.

I walked into the back of his shop where radios, tubes, and other components were spread out on the work benches alongside the various tools of the trade. And then I saw it ... a magnificent looking thing in a brown cabinet with a round screen about 12 inches in diameter. As Mr. Barrett had said, it wasn't hooked up, but that didn't matter. I walked over, placed my hand on the screen, and stared at it. In my mind it wasn't blank. I was watching Stan Musial at the plate, hunched over in his unique batting stance, waiting for a pitch from Warren Spahn. Imagine, watching a baseball game in your own living room!

"What do you think?" Mr. Barrett asked.

"Is something wrong with it?"

"Nothing. The people who own it just moved here from St. Louis, and they thought it was broken. They just don't realize how far away we are from the signal."

The first time I ever actually saw a TV in operation was about two years later at a neighbor's house. Fred Matthews, a good friend of my father, bought one and put it in their parlor, then lined up chairs as if it were a theater. I lived for those times when they would call and say, "TV is pretty clear tonight."

We would join others who had been invited and watch whatever offering KSD-TV in St. Louis or WMC-TV in Memphis had to offer. And that was part of the problem. Sikeston is exactly halfway between the two cities, and both stations broadcast on channel five. Often—especially on days when the signals were strong—the two stations would compete with each other, one picture swiping across the other. When that happened, you would have to turn the outside antenna, by a remote electric device, from north to south or vice versa. Often you would be following two different shows at the same time.

We had to have our own television. But my dad would always say, "Not until the screen gets a little bigger and a TV station comes close enough to actually watch."

I was so convinced that our family was the last in Sikeston to get a TV set that I used to give directions by saying, "It's easy to find us. We're the only house on the block without a TV antenna."

Then both of my dad's requirements happened at the same time. We got a TV station only 32 miles away, and Stromberg-Carlson built a 21-inch set with an outwardly curved screen. "That way you can see it no matter where you are in the room," my dad said proudly when he announced he had bought one.

I can still remember the excitement of having the TV set delivered and the antenna mounted on the roof—absolute evidence to anyone who drove by that *this family has a TV set!*

The new station was on channel 12, which protected it from any interference from a competing station on the same channel. It didn't start broadcasting until late in the afternoon, opening with the National Anthem, then the Circle 12 Corral, which showed old Western movies. And now, with a clear picture at last, we could really enjoy television: Jack Benny, Ed Sullivan, I Love Lucy, Red Skelton, Jackie Gleason.

After the 10 o'clock news, a disembodied voice would announce: "This concludes the day's broadcasting for KFVS-TV, Channel 12, in Cape Girardeau, Missouri." Again the National Anthem would play, with jet aircraft flying across the screen against a waving flag.

The set was not yet turned off. We would watch until the last note was played, then the screen would be filled with a test pattern consisting of a lot of concentric circles of broken lines, horizontal and vertical bars, all centered around the face of an Indian chief.

My Grandpa in an Old West Shootout

"You better be totin' a gun next time I see you, 'cause I'm gonna shoot you, you son of a bitch."

You might think that's a line of dialogue from one of my books, but you'd be wrong. That deadly threat was made against my grandpa, well over 100 years ago.

First, let me tell you a little about my grandpa. He may well have been the most stoic person I ever met, and even before I knew the meaning of the word, I understood the personality. Pop, as we called him, would listen to the Grand Ole Opry when there wasn't too much static, but that was virtually the only thing he ever listened to on the radio. Every time he bought a new car, it was exactly like the one he had been driving—a gray, four-door Chevrolet. He didn't dicker over the cost, he didn't examine other cars for interest or bargain, he just went out and bought a gray, four-door Chevy. Then, like as not, he would pick up a plow he was having repaired, throw it in back of the new car, perhaps ripping up the seat, then head back to the farm, sometimes driving across a field to deliver the plow.

He chewed tobacco and spit it out the window, so that by the time he got the new car home, the driver's side would be stained with tobacco juice and the whole vehicle would be so dirty from its trip through the field that nobody would realize it was new.

Another thing about Pop's car is that the back floorboard would be filled, up to level with the back seat, with shotgun shells. As kids, when we sat in back, our legs were stretched out in front of us, resting on that bed of shotgun shells, which was replenished every time he went to town. Pop was a very good shot, he carried his shotgun in the car and would often shoot a flying dove from the moving vehicle.

"Boy, go get that bird," he would say, stopping the car long enough for us to recover it.

But, back to the threat made against him. Pop was a young man when it happened, and he had been hired by a store owner to take a cow from a farmer who owed them money. No legal confiscation, no court order, just a sense of taking a man's cow, because the man owed the store money.

Pop took the cow into town, and when the farmer saw him later, he issued that challenge for him to be "totin' a gun next time I see you."

A few days later, Pop was riding his mule down a rural road, when he saw the man who had threatened him. The man shot at Pop, who shot back, knocking the man off his horse. He crawled out into a cornfield, but Pop didn't go after him. Instead, he went to the sheriff to tell him what happened.

The man didn't die. Pop was tried for the shooting, but there were witnesses who had heard the man threaten my grandpa, so he was found innocent.

Now, let's forward to 1947. We got word that the same man, who no longer lived in the area, was coming to see Pop. My two brothers, my cousin, and I hid behind the corner of the house, awaiting what we expected would be another shooting. But when the man arrived, he and my grandpa shook hands, talked for a while, shared a few laughs, then the man got back in his car and drove away.

Our hearts sank. We desperately wanted to see an Old West shootout.

The Big Game

In the days before Little League baseball and football, we played our own games, without coaching, without referees, and without any kind of organization. Baseball games were held in empty lots, which resulted in a lot of broken windows—I think my personal count was five—and broken asbestos shingles. Remember, this was in the '40s, before the asbestos scare, when the inside walls of schools were coated with asbestos and half the houses had asbestos siding.

On Greer Street we played football in back yards or between houses, until Mr. Matthews harvested his soybeans in the five-acre field across the street from my house. Then we would very methodically pull up enough of the remaining plant stubs to make a field where we wouldn't get our eyes poked by a stick anytime we got tackled.

We had a field, and we broke it in by playing pick-up games among ourselves. But the idea began to grow for a real football team—the Greer Street Tornadoes. We assembled our roster, but we needed another team to play. At school, I asked Alfred Sikes to put together a team from his side of town. Arrangements were made and the date set for the next Saturday morning. It wasn't hard to fit the big game into our schedule, since they were the only other team we knew. We only had nine players, not enough for an eleven-man team: Neil and David Lewis, Gerald and Ray Kelly, my brother Tommy and me, Kenny Kern, Danny Dunnigan, and Earl Ray Underhill. Since five of us lived on Greer and the football field was there, we named ourselves the Greer Street Tornadoes. I'm not sure I remember how we chose that name.

Every one of us had shoulder pads and a helmet, compliments of the Sears, Roebuck and Co. Catalogue, which was the Amazon of the day. And we had a secret weapon: uniforms! Well, they weren't really uniforms, but at least we were uniform: blue jeans and white sweatshirts. We were a little concerned about the colors. Sikeston High School was red and black, and rival Charleston High School was blue and white. I hated that we looked like Charleston, but we had to go with what we had.

The north end of town, home of our competitors, was quite a distance away. But in those days, once you had a bicycle, the entire town of Sikeston was open to you, so we knew they'd have no trouble getting here. Then, at the appointed hour, 10 bicycles arrived. *The game was on!*

We had a quick preliminary discussion about the fact that they had ten players and we had only nine, but we struck an agreement that they would only play nine at a time. But then we realized there was a bigger problem. They also had uniforms: blue jeans and white sweatshirts!

It really wasn't that big a deal. We all knew each other, so we wouldn't be throwing the football to the wrong receiver. We would just throw it to Tommy or Ray or David or anyone else on our team. We didn't bother with such things as "ineligible receivers." We threw the ball to anyone who was open.

We also didn't have a system of measuring downs, so there were no first down after advancing 10 yards. You either scored in three downs or you punted. Therefore, there were no long drives down the field. Touchdowns, when they occurred, would be a completed pass or a run to the goal.

The field was right across from the Nailing Truck and Tractor Company, and five or six of their workers came over to watch us play. *We had spectators!*

A few of the players had watches, so we kept a running clock, playing 15 minutes per quarter, and we actually changed ends at each quarter.

I honestly don't remember who won this first game. After all, it was played over 70 years ago. I just know that it was a lot of fun, nobody got mad, nobody got their feelings hurt because they lost, and any injuries we may have suffered weren't enough to make us quit playing.

A Terrible Tragedy

Sometime between Christmas and New Years in 1949, some friends and I were playing football in David and Neil's back yard when it started raining, so we took refuge in the garage. We heard the roar of aircraft engines overhead, which wasn't unusual, as we lived very close to the airport. But we had never heard anything like this, so we ran outside and looked up to see an early-design twin-engine Cessna called the Bamboo Bomber, because the wings were made of wood.

The airplane was frighteningly low and made a very steep climb, which was the pilot's mistake. He stalled at the top of the climb, and the plane went roaring straight down, crashing in an open field. There were two couples aboard, and all were killed in the. We later learned that they were from Chicago, en route to New Orleans to attend the Sugar Bowl. As eye witnesses, we were later interviewed by the FAA.

You'd expect witnessing something so horrendous might have dampened my enthusiasm for flight, but somehow it piqued my fascination even more.

BB Gun Wars

When I see kids getting into trouble for making their hand into the shape of a gun, or drawing a picture of a gun in school, or carrying a knife in their fishing kit, I marvel at how much things have changed since I was a kid.

In the 1940s in Sikeston, a Daisy Red Ryder BB gun cost $3.75. My brother Tommy and I picked cotton to earn enough money to buy our BB guns. Picking cotton paid 3.5 cents a pound, and it was easy to get a job. You just rode your bike to a field where they were picking, parked your bike alongside the field (we didn't have to worry about it being stolen in those days), found an empty row, and started picking. You had to have your own toe sack, known to you Northerners as a burlap bag. The toe sack could hold about 10 pounds, worth 35 cents, and that was about as much as you could pick from the time school was out until they quit.

When weighing at the end of the day, they would subtract one pound for the sack, so we kids talked about putting a line of fishing weights in the bottom of the sack, but none of us ever did.

Once we had enough money for the BB guns, we became fierce hunters. The other kids who lived in the same block—Gerald and Ray Kelly, David and Neil Lewis, Earl Ray Underhill, and Kenny Kern—also had BB guns, and we patrolled the area, hunting birds as if they were foreign spies and shooting them out of trees and off telephone wires. These weren't game birds but were mostly sparrows and blackbirds. We never shot a robin or a dove. We would fill our pockets with the birds we shot, compare notes as to who got the most, then feed the birds to the neighborhood cats. There were always dozens of uncollared and unclaimed cats running around.

We also played war or cowboys with our BB guns, actually shooting at each other, the only rule being not to aim at the head because all our mothers had warned us we might "shoot an eye out."

Roy Beck, who knew everything there was to know because he was already in high school, told us that "the BB doesn't get to full speed until it's about ten feet away from the gun." He insisted that if you would hold your hand over the end of the barrel and pull the trigger, it wouldn't hurt. I was stupid enough to put his theory to a test. His theory was wrong. It hurt like hell and I had a bruised palm for a couple of weeks.

Willeana Piercie lived behind our house. She was a mean old lady who would take a ball if it went into her garden. One day while shooting our guns, a BB bounced off her garage window. To our surprise, the glass didn't shatter.

"Look at that," Gerald said. "The window didn't break."

The original shot had been an accident, but curious why it didn't break, Gerald shot it again. It still didn't break, so I shot it. It still didn't break.

"Boys, you know what that is?" Neil asked in a knowing voice. "That's bulletproof glass."

"Huh, uh. Why would she have bulletproof glass?"

"Because she's a mean old lady, and she's afraid someone'll shoot her. That glass can't be broken."

"Sure it can," I insisted.

"I'll bet you a dollar it can't be broken," Neil said.

"A dollar? You bet a dollar?"

"Yes."

A dollar was a lot of money—three days of picking cotton. I picked up a heavy chunk of firewood. "A dollar?" I asked again, to be sure of the terms.

"A dollar," Neil insisted.

I threw the firewood at the window, and it came crashing down with a noise that could be heard all over the neighborhood.

Neil, David, Gerald, Ray, Tommy and I ran, but not any farther than my back yard, directly across the alley from Mrs. Piercie's house. We all assumed the same position, on our stomachs with our arms covering our head.

Mrs. Piercie came out of her house, screaming bloody murder. My mother also came out, and the two of them stood in our back yard over the prostrate forms of six young boys.

"What is it?" my mother asked. "What happened?"

"I'll tell you what happened," Mrs. Piercie exclaimed. "These brats shot out the window in my garage."

"Which one of you did it?" Mother asked.

After a moment's hesitation, Neil jumped up. He was the oldest and was going to take responsibility, since he was the one who made the bet. Good for you, Neil, I thought. I'll owe you one.

"He did it!" Neil shouted, pointing at me. "Dickey did it, Dickey did it!"

"Dickey, did you shoot out Mrs. Piercie's window?"

"No, ma'am."

"Dickey's telling the truth, Mother," Tommy said.

Good for Tommy. He was my brother, after all, and was taking up for me.

"He didn't shoot out the window. He threw a log through it."

I got a whipping for it. But if my memory serves me right, when Mrs. Lewis found out what happened, Neil got a whipping, too. David, Gerald, Ray, and my brother Tommy got away clean.

Those who have left

When I was very young, I thought the entire world consisted of Sikeston, Missouri, at one end and Jackson, Mississippi, at the other, separated by very long train or car trips. But the trip to Jackson was always worth the long ride, because my cousin Ronnie lived there.

Ronnie was a year younger than me and a year older than my brother Tommy. My mother and my Aunt Evelyn once made identical outfits for us to wear, and I remember someone asking if we were triplets. When they learned we weren't, they said we must be the Three Musketeers. That sounded good to us, so for a while that's exactly what we called ourselves.

I'm the only one left of the Three Musketeers: my brother Tommy died 30 years ago, and my cousin Ronnie died about 5 years ago. My best friend died 10 years ago, and high school classmates, boys I played with, girls I dated, lifelong friends, and family members are all dying. I started to type "are leaving me," but they aren't really leaving me. I have but to think of them, to remember a pleasant occasion with them, and they are still with me.

At this very moment, I've no doubt that Ronnie, Tommy, and I, wearing those identical outfits, are at Monkey Island at the Livingston Park Zoo in Jackson. Ronnie just threw some "Boston Beans" over the wall, and we're laughing as the monkeys scramble to get them.

Good for Her

When I was a schoolboy back in the forties and fifties, there was a girl in my class, Lillian S., who was by far the smartest of any of us. In all her classes, she received an E, equivalent to an A today (back then, Missouri's grading system used E-S-M-I-F instead of A-B-C-D-F). But it wasn't just Lillian's grades but her general knowledge that set her apart. I've always considered myself well-read, and I haunted the school libraries, but Lillian knew history that had not yet been taught, she knew science that was beyond our curriculum, and she was always up-to-date on current events.

Back then, the eighth grade was located in the high school building, and after the eighth grade we could take some elective classes the following year. I remember how excited we were when signing up for freshman classes, but Lillian didn't sign up for anything. We were shocked when she told us she would be leaving school after completing the eighth grade, the earliest at which you could drop out of school. But Lillian, the smartest one in our class? Why would she do such a thing?

When we questioned her, she told us tearfully that her father felt that the eighth grade was as far as any girl should go, and if Lillian was going to live in his house, she would have drop out now and get a job. I can still remember the sense of unfairness I felt then. No, it was more than unfair, it was an outrage, and I wasn't the only student who felt that way. Many of my classmates expressed the same surprise, sorrow, pity, and anger over what Lillian's father was making her do.

We came back the next year, hoping to see Lillian, but she wasn't there. Finally we accepted what had happened, and we eventually graduated and went about our lives. But for many years afterward, I would sometimes find myself wondering what had become of Lillian. Then, out of the blue about 20 years ago, I got a telephone call.

"Mr. Vaughan?"

"Yes."

"I've read many of your books and wanted to talk to you."

"Thank you," I said courteously to this apparent fan.

"Are you the Dickey Vaughan who went to school in Sikeston?"

I smiled. People who call me Dickey have known me for a very long time.

"Yes, I am."

"We were classmates once, though I didn't graduate with the class. My name is Lillian S. Do you remember me?"

"Yes!" I practically shouted. "I can't tell you how great it is to hear from you! Whatever happened to you, Lillian? Where are you?"

"Well, after I left home, I was able to go back to school. I was older than the other students, but it was a private school so there wasn't that much made of it. After I graduated from high school I passed the entrance exams and went to college. Now I have my PhD and taught English at Washington University in St. Louis until I retired. I just wanted to reach out and talk to someone that I knew from a long time ago."

I filled Lillian in on what I knew about all of our classmates, and our conversation lasted for nearly an hour. I haven't heard from her since, but there is no need for me to. One of the mysteries of my life has been solved, and it could not have had a more wonderful outcome.

Harry Truman as my Debating Partner

I was a high school debater and enjoyed it very much. In the first debate of my career, my partner and I, both freshmen, debated a couple of seniors from Tilghman High School in Paducah, Kentucky. We lost that debate . . . but in four more years of debating, I never lost again, and I wound up my debating career as president of the Debating Club.

Harry S. Truman was president in my first year of debating, and I wrote him a letter asking for his opinion on the debate question that year. I got a response, but it wasn't quite what I was looking for.

> Dear Dickey Vaughan:
> The President is too busy with world affairs to give you an answer to your question, but he wishes to congratulate you for your interest in debating, and he wishes you the best of luck. Here is the President's autograph for you.
> *Matthew J. Connelly*
> *Appointments Secretary*

Below Connelly's name was President Truman's signature.

Initially, I was very disappointed. This letter wasn't going to do me any good in debates. *But,* the letter came in a franked envelope with Truman's signature where the three-cent stamp would have been. (Yes, three cents. This was 1952, remember.) And that opened up a world of possibilities.

I retyped the letter.

> *Dear Dickey Vaughan:*
>
> *It is very good to see that a young man like you is so interested in the subject of Universal Conscription for all 18-year-old males. Like you, I believe that everyone owes a debt of service to his country, whether it is to be served as two years in the military or a two-year obligation working for the government in some other capacity. Exemption from such service should be possible, but such exemptions should be in keeping with military exemptions: physical condition, academic exemptions, or unique family circumstances.*
>
> *Please feel free to use this letter in all your high school debates, and I wish you the best of luck in supporting this important question.*
>
> *Harry S. Truman*
> *President of the United States.*

You might notice, I skipped completely over Connelly's signature and went right to the big man himself. I now had a letter that I could use to my advantage, in any way I wanted.

I didn't use the letter in every debate, but I used it frequently, validating it by showing the opposition and the debate judge not the letter itself, which I explained I had "retyped in order to preserve the original," but the franked envelope. In each case, I would read the letter to great fanfare...and, I used that same ploy for the next four years, dispatching teams from St. Louis, Cape Girardeau, Poplar Bluff, Kennett, and with particular joy, Paducah.

As it turned out, even though Truman was no longer president in my sophomore year, he was opposed to the allocation of industrial capabilities among allied nations. The following year, he was for the defense of the islands of Quemoy and Matsu from the Chinese Communists, and by my senior year he was opposed to abandoning the Electoral College. Actually, in the latter, he was both opposed and for it, as there were times during that year that I debated both sides.

Tommy Murchison was my senior-year debating partner, and he lived in absolute fear that one of the debate judges would discover what I was doing. But by the end of the fourth year, I was using Truman's letter—or I should say letters—with such confidence that I was not concerned about the judges. By that time I had nearly convinced myself they were authentic.

Many years later, when I worked for Congressman Tommy Downing of the First District of Virginia, I shared that story with him. He laughed and said that with my ability to pull that off, I should go into politics. Today, I fear that too many of our politicians may have achieved their current status in much that same way.

An Early Military Mishap

I was 15 when I joined the National Guard, but I wasn't the only one who was underage. Our company commander, who was also a neighbor, was pretty lenient about underage enlistments, perhaps because his own two sons were among the group.

This was during the Korean War, and as I look back on it now, I realize how fortunate we were that, unlike WWII and recent conflicts, the National Guard was never mobilized. I enjoyed my time in the National Guard, and it served me well during my subsequent service in the U.S. Army—including the added benefit that when I entered the army I already had 3 years of longevity for pay. And unlike the others in my basic training company, I held the rank of private E-2, not E-1.

Our sergeant in the National Guard was a man named Miner D. Cobb. First Sergeant Cobb was a very good soldier who gave no quarter.

My first summer, we went by train to Camp McCoy, Wisconsin, for our two weeks of encampment. When we arrived, Sergeant Cobb assigned me as driver of the colonel's jeep.

"First Sergeant, are you sure you want me to do this? Wouldn't it be better to have someone who is . . . uh . . . a bit older?" (I didn't dare tell him I was only 15.)

"I don't care how old you are, Vaughan. You are a member of my company, and you will perform your duty just like everyone else."

"Yes, First Sergeant."

I could drive—in fact, I had been driving since I was 12 years old. In the early 1950s, most of us started driving early, and since I had never even been in a car with automatic transmission, shifting gears was no problem. I drove the jeep to the battalion HQ and reported for duty.

The battalion commanding officer was Colonel Wickham, who was also a member of the Missouri Highway Patrol.

"How old are you, soldier?" Colonel Wickham asked.

"I'm seventeen, sir."

"Let me see your driver's license."

"Uh, I didn't bring it with me."

"I'm not going to let anyone drive me around who doesn't have a driver's license." He looked at his watch. "It's too late to get another driver. Scoot over, I'll drive. You'll stay with the jeep and keep an eye on it."

"Yes, sir," I said. Getting out, I started to hop into the back.

"No!" Colonel Wickham said. "You ride up front. It's bad enough that a private has a colonel as his personal driver . . . I'll be damned if you're going to ride in the back."

I rode in the front seat for the rest of that day, sitting in the jeep while the colonel went from place to place to take care of whatever business it was that colonels take care of. I sat outside the officer's club during lunch, but he brought me a hamburger and coke when he came back.

It was after dark before the colonel's day was finished. He stopped along the edge of some trees and pointed. "Your company is bivouacked just on the other side of these trees, no more than half a mile. Tell Captain Lewis to send me a new driver tomorrow—one with a driver's license."

"Yes, sir," I said.

The colonel drove away, and I started into the woods. It was already dark, but it got much darker in the woods, with a lot of strange sounds.

Are there bears in Wisconsin? I wondered.
What was that? That sounded like a panther!
The mosquitoes are as big as H-13 helicopters.
I thought the colonel said it was only half a mile.

I wandered around in the woods for at least three hours. I was tired, mosquito-bitten, and scared. "Hey! Does anybody hear me? I'm lost!"

I knew everyone was going to tease me, but at that point I didn't care. It was after midnight, and I was "this close" to calling for my mother—and if we had cellphones then, I would have. Finally, sheer exhaustion caused me to sit down under a tree. The mosquitoes continued to torment me, and the wild beasts—most of which sounded rather like frogs and crickets—continued to haunt me.

"*Vaughan? Private Vaughan? Vaughan?*"

Someone was shouting my name. Not just someone, but several people. The shouts had awakened me. It was morning now, and shafts of sunlight were stabbing through the trees. The bears and panthers were gone.

"Maybe he went AWOL, Sarge," I heard someone say.

"I'm here!" I called. "I'm here!"

The entire company had turned out to search for me, costing almost half a day of training, First Sergeant Cobb was quick to point out.

Years later, I was a chief warrant officer in the 7th Cavalry, pulling duty officer, when my duty sergeant reported to me. He was none other than the same man who had been my sergeant back in the 140th Infantry Regiment of the 35th National Guard Division.

"Hello, Sergeant. Do you remember me?" I asked, hoping

the fact that I now outranked him would make him a little nervous.

He smiled. "I remember you, Mr. Vaughan. You're the doofus who got lost in the woods, aren't you, sir?"

So much for intimidating a good soldier.

High School Band

One of my most pleasant memories is my time in the Sikeston High School band. In the early 1950s, the eighth grade was housed in the same building as the high school, so eighth graders were allowed to join the band. I had studied the trumpet since the third grade, seeing myself as a future Harry James . . . who, by the way, was married to Betty Grable, so why shouldn't I have such aspirations?

"Do you want to be in the band?" Mr. Keith Collins asked when I first showed up at practice.

"Well, yes, sir, that's sort of what I had in mind," I replied, wondering why he had even asked.

"Good. I have too many cornet players who are better than you, so you are going to have to switch from the cornet to the French horn."

The French horn? I had never even heard of a French horn. But I wanted to be in the band, so I agreed.

When we began to get ready for the marching band season, I started going through my music.

"You don't need music," Eddie Webb told me. Eddie was

a junior who sat beside me.

"What do you mean, I don't need music?"

"Listen to the bass horn. When the bass goes *boom*, you go *deet*. Marches are like this: *Boom deet, boom deet, boom deet, deet*. We're the *deet*. Do it in middle C. That's all you have to know."

Well, there was very little musical about that, but it did make playing march music very easy.

The band trips were fun, especially the ones where you were gone all day, such as the Soybean Festival in Portageville, the Cotton-Picking Contest in Blytheville, and the Band Festival in Jackson. And, as I got older, I discovered a secret that only those of us in the band knew: If you sat next to a girl on the bus, you smooched the whole trip back. It didn't matter if she was going steady or even who she was going steady with. When I was a freshman, the steady girlfriend of the starting quarterback, who was a senior, took great delight in teaching me how to kiss. Is it any wonder that now, 65 years later, I can still remember that?

I liked the concert season the most, though, because then the French horns actually did contribute to the music, much more than *boom deet, boom deet, boom deet, deet*. And, it introduced me to classical music, which is still a huge part of my life. I listen to it as I write. Mr. Collins always let the seniors have a "moment" during the concert season, and I got to play a passage from Ravel's *Pavane For a Dead Princess*. The key valves on a French horn were operated by strings, and just before I stood for my 12-measure passage, the strings broke. When I stood to play, rather than holding the French horn as it should be, with the right hand in the bell of the horn, my right hand was wrapped around the valves. Mr. Collins knew exactly what happened . . . and today I can still remember the expression of absolute horror on his face. After the concert, he "suggested" that perhaps I should have checked the strings beforehand—something I had realized the moment the strings broke.

On the night of graduation, the band chairs of the graduating seniors were set up but left empty. It was a fitting tribute. There are people we encounter along the way whose impact on our lives last a long time. For me, and I know for many others who were in the band under him, Mr. Keith Collins was one of those people.

Lord Help the Mister that Comes Between Judy and her Sister

Judy was a very pretty girl, and when we were dating, I always felt proud to be seen with her. She was one grade behind me, which made our ages compatible, as I was younger than any girl in my class. Judy had a younger sister, Pat, who was just as attractive and an outrageous flirt.

Judy was preparing to go to a band camp, so on the day she was to leave, I went over to her house and asked Pat if she would like to go out with me. It was then I learned that Judy had not yet left for band camp. Pat turned me down . . . and so did Judy. My brief school romance was over.

Lord Help the Mister that Comes Between Judy and her Sister

Army: Enlisted Man

Joining the Army

"Don't take more than a couple of dollars with you, and don't take any clothes other than what you'll be wearing," the recruiting sergeant told me. "You'll have to send all your civvies back, because you can't have them while you're in basic training."

"All right," I agreed. "What next?"

"You'll report to the processing center in St. Louis. There, you'll be sworn in, then put on a bus to Fort Leonard Wood." The sergeant smiled. "You'll be on the government payroll then. Actually, you already are. Here's a travel voucher for a Greyhound bus ticket to St. Louis. You'll leave at one o'clock on Friday."

"Friday?" I was a little disappointed, I wanted one last weekend to spend with my friends to tell them that I was going into the army.

"The sooner you start, the sooner your three years will be up."

"All right, that makes sense."

My mother took me to catch the bus, which at that time

was next to the Rustic Rock restaurant. Embarrassed that my mother was taking me to the army, I asked her to say goodbye in the car, let me out, and go home. "I'm a soldier now."

"You're seventeen."

"I'm a seventeen-year-old soldier."

I looked around to make certain nobody was watching, then I let her kiss me goodbye, then watched her drive away. Later though, when I got on the bus and looked through the window, I saw that she hadn't actually left. The car was in front of the Rustic Rock, and Mother was wiping her eyes with a tissue.

Damn, that made me choke up a bit, and I had been fine until then.

The recruiting sergeant told me to sit in the front seat and tell the bus driver where I was going, because he could let me out one block from the processing center. It would be easy, because I had no luggage.

"I've taken a lot of boys up," the driver said. "Don't normally take them on a Friday, though."

When we reached St. Louis, he stopped the bus, opened the door, and pointed. "It's one block down there on the left. There's a big sign out front. You can't miss it."

Excited and curious, I thanked the driver, then walked down to the processing center. When I tried to open the door, though, it was locked. I banged several times, and a corporal finally came to answer it.

"What do you want?"

"I'm supposed to be here to be processed."

"Not today you ain't. We're closed until Monday."

"Closed? How can the army be closed?"

"Come back Monday."

"What am I supposed to do until Monday? I don't have any money. I don't have anywhere to go."

"You ain't in the army yet, so you're not the army's problem." The corporal closed the door, and I heard it lock.

I walked back a few steps and stood on the street, angry

and a little frightened. What would I do? It was early March, and it was cold. I saw a telephone on the corner, so I decided to call my mother. After all, I wasn't in the army yet, and like she said, I was only 17.

"I don't know what to do," I said after she agreed to accept the charges.

"The Hegemans live in St. Louis now," Mother said, referring to our former next-door neighbors. "We still keep in touch; I'll call them. What is the telephone number there? I'll call you back."

I gave her the number, then stood waiting inside the telephone booth, partly to be out of the cold wind, and partly to will the phone to ring. What if she couldn't reach them? What if she could? What could they do?

The Hegemans, I remembered, had a daughter, Diane, who was 12 years old at the time and somewhat of a pest.

The phone rang, and I answered. "Dickey?" It was Mrs. Hegeman.

"Yes!"

After a brief conversation, I gave her the address, and within half an hour they showed up in a 1955 Ford. Mr. Hegeman rolled the window down and smiled at me. I had never been happier to see anyone.

"Get in," he said as the back door opened.

"Hi, Dickey. Do you remember me?"

The girl who spoke was about 15 and no longer the gangly pest I remembered.

"Diane!" I said.

"We're going to go have a pizza pie," she said, enthusiastically.

At that time, I had never heard of pizza and while I thought a pie might be good, I was hungry for more than just a piece. Maybe it would be apple, with melted cheese on top.

The place we went to was in the basement of what appeared to be an ordinary business building. Inside were several tables, with red-and-white checkered tablecloths. It was then that I learned what a pizza pie really was, and to be honest, my first reaction to it wasn't all that favorable. But I was hungry, and they had bought the meal for me, so I tried to generate as much enthusiasm as Diane, who actually seemed to be enjoying it.

The next night, Diane and I went to the movies together. It turned out to be a very pleasant weekend, and when Mr. Hegeman took me to the processing center on Monday morning, I was rather glad that I had one last weekend as a civilian.

I was processed in on Monday, and when they learned I was in the National Guard, the sergeant asked what my job was.

I was an M209 code convertor operator, and when we were deciphering codes during National Guard meetings, we were in a private room because the instrument was classified. We sent and decoded notes to each other, such as: *Ejp fp upi yjoml om itryyort smm pt dir?* (Who do you think is prettier, Ann or Sue?)

"I can't tell you what my job is. It's classified. I need someone with a secret clearance."

The sergeant called the lieutenant over. "I have a secret clearance," the lieutenant said. "What was your job in the National Guard?"

"I was a . . ." and here I halted, not sure who was listening in. "M209 code convertor operator."

"Sergeant, what in heaven's name is an M209 code convertor?"

The sergeant laughed. "Those things went out during World War II."

By the end of the day I was in the army, and according to the sergeant who informed us that we now had a $10,000 army insurance policy, we were now worth more to our family dead than alive.

It would be 23 years, tours in Korea and Germany, and three deployments to Vietnam, before I would be a civilian again.

Basic Training

When I was in Basic Training at Fort Leonard Wood, we were having a "GI party." No, we weren't sitting around drinking beer, eating dips, and listening to Bill Haley and the Comets or Elvis Presley. A "GI party" meant we were cleaning the barracks, which in those days were the old WWII wooden barracks with 32 bunks, eight double-stacked on either side.

Under the watchful eye of our platoon sergeant, we were scrubbing the floor with bristle brushes and lying about our high school exploits. I had "gone steady" with Miss Sikeston, I bragged, but let her down easy when I joined the army. That was partly true. Judy had been Miss Sikeston, but she dumped me long before I joined the army. Also, I had "just missed being valedictorian," and "scored winning touchdowns against Cape Girardeau, Jackson, Poplar Bluff, and Charleston."

Then, when I thought perhaps I had laid it on a little too thick, I decided to humble myself by telling the story of how Miss Jones kicked me out of the play. "She was awful!" I said. "She was the worst teacher I ever had! I don't know how she ever got to be a teacher!"

I continued to berate Miss Jones, adding as many negative things as I could come up with. Then our platoon sergeant, who hadn't said anything beyond, "you missed over there," spoke up. "Vaughan, where did you say you went to high school?"

"Sikeston High School, Sergeant."

"And what's the name of this schoolteacher you are talking about?"

"Her name is Miss Jones. She's a redhead, and everything they've ever said about redheads is true: She is a hellion." I added a few other things to the delight of the other soldiers.

"Vaughan, I've been wondering who I would pick to clean the toilet bowls and the urinals, and I think you have just volunteered."

"What? How? Why me?"

"Let's just say I don't like the way to talk about schoolteachers."

"Ha! You wouldn't say that if you knew Miss Jones," I said with a smug grin.

"Oh, I know Martha Howard Jones, all right."

I felt a little queasy. I hadn't, at any time in my tirade, given Miss Jones's full name. How did he know it?

"I think you had better get started cleaning in the latrine, now."

"Yes, Sergeant Jones."

Well, come on, Jones is a very common name, isn't it? What are the odds of a connection here?

When I went home on leave after basic training, I stopped by Miss Jones' house. She greeted me most graciously, invited me in and offered me cookies, then laughed with me about some of the things that had happened while I was in school. She had been my sophomore English teacher, and because she was the sponsor of the drama club, had directed me in other plays. But before I could even ask the question, she answered it for me.

"My brother, Johnny, told me he had you in his platoon. He said you were a good soldier and must have been a brilliant student and a great football player."

"Uh . . . Miss Jones, I . . . uh . . . want you to know that Sergeant Jones was the best NCO in the whole company."

"Like you were the valedictorian and single-handedly won every football game?"

"I . . . uh . . . well, sometimes I start talking and maybe go a little overboard."

Miss Jones laughed. "Don't worry, Dickey, I didn't disabuse him of any of your imaginary accomplishments. One of the things I liked about you was your creative mind . . . and if it helped you get through basic training, so be it. Johnny has told me how hard it is on you boys up there."

Turns out, I was wrong about Miss Jones. She was a good teacher and a class act. I wasn't really that crazy about her brother, though. I mean, someone of my sensitivities, cleaning latrines . . .

A Very Brief Romance

It was September 1956. I was 18 years old and had just graduated from an aircraft maintenance course at Fort Rucker, Alabama, and was being kept on as an instructor. I was given a two-week leave, so I took a Greyhound bus that rolled out of Ozark at about five o'clock in the afternoon en route to Sikeston, Missouri. It was after dark when the bus stopped in Montgomery, and a young, very pretty girl boarded the bus and asked if she could sit beside me. Smiling, I said, "Yes," as I adjusted my tie, pulled down the "Ike jacket" of my uniform, made certain my Expert Marksmanship badge could be seen, and checked the spit-shine on my brown shoes.

She told me that she was 18 years old and on her way to Memphis to visit her grandparents.

"I'll be going through Memphis, as well," I said. "According to the schedule, we're supposed to arrive there at oh three hundred hours." I used the "army lingo" to impress her, and it worked.

"What time is oh three hundred?" she asked, clearly fascinated.

"Three o'clock in the morning."

"My," she said, her smile broadening. "We'll be sitting together for a long time."

Somewhere around Prattville, the conversation stopped, and the kissing started. We smooched all the way from Prattville to Memphis, not even bothering to get off the bus when it made its few stops. At Memphis she got up and told me goodbye. I looked out through the window, hoping she would look back and wave, but she didn't. We never even told each other our names.

Somewhere in America today there is an an 83-year-old woman, no doubt a grandmother many times over, perhaps even a great-grandmother. I hope she has had a great life, with a wonderful marriage and terrific children, grandchildren, and great-grandchildren. But I wonder if, from time to time during her life, she ever thinks of the "handsome" soldier boy that she smooched with for 325 miles, back when she was a young, very pretty, 18-year-old girl.

A Letter Home

When I went to Fort Rucker in 1956 to attend crew chief school, we were in the same barracks the WWII soldiers had used, and on the wall in the furnace room was "Instructions for Firemen" with a posting date of 1943. In the latrine there was a small slit beneath the mirror over the lavatories with the notice, "Drop used razor blades here." I used to wonder how many blades had been dropped into that wall over the years and think about the men who had dropped them.

After I completed crew chief school I was retained as an instructor in the same course I had just completed. I have a confession to make. That was the best assignment for me. I'm a good speaker and made a pretty good instructor, but I wasn't very good in the "hands on" aspect of actual maintenance operations. The old adage of "those who can, do, and those who can't, teach," was never truer than in my case.

Our company commander was a "mustang"—a man who had come up through the ranks. Such a thing isn't even possible now, but Captain Poppell didn't even have a high-school diploma. He once said that during the Korean War he had gone before a promotion board with three other sergeants first class, hoping to make master sergeant. He didn't make it but was given the rank of second lieutenant as consolation.

"I was disappointed," he said. "I wanted that other stripe."

Captain Poppell had what is called a command voice, and because of that, he often took charge of the monthly retirement parade ceremonies, giving orders in a stentorian voice that rolled out across the parade ground and was easily heard by the guests in the bleachers.

"*Riiiight!*" he shouted one Saturday morning in such a voice. Then, in the same tone, volume, and pace, he corrected himself: "Dammit, I mean left face!"

Something you never wanted to do as a low-ranking enlisted man was to be summoned before the company commander, but one day just such a summons came.

"Vaughan, the CO wants to see you," said Corporal Cates, one of the company clerks.

"Why?"

"How'm I supposed to know? I'm just a corporal. The ole' man says jump, I don't even ask how high, I just jump."

Filled with trepidation and some curiosity, I walked down to the orderly room. I wasn't aware of having committed any transgression, but you never knew.

"Private Vaughan," Captain Poppell said after I reported to him. "When is the last time you wrote to your mama?"

"Sir?"

"I've got a letter from your mama," the captain said. "Let me read something from it."

Clearing his throat, he began to read in the same voice in which he gave the parade ground commands, a voice that carried to everyone in the orderly room, as well as the adjacent arms and supply rooms.

"I am so concerned about my son, Dickey." Captain Poppell looked up. "Would that be you? Does your mama call you Dickey?"

"Yes, sir," I admitted as quietly as I could.

Poppell continued to read. "I have not had a letter from him in two weeks. I do hope that he hasn't been injured in some way, and the army just hasn't told me."

Poppell lay the letter down on his desk. "You go out into the clerks' office and write your mama a letter, then you bring it in here so I can read it. And if you dare complain to your sweet mama about her writing me, I will make your life miserable. Do I make myself clear, soldier?"

"Yes, sir."

A moment later I was sitting in the clerks' office of the orderly room, writing my mother.

"You be sure and write a nice letter now, Dickey," Corporal Cates said, and the others in the room laughed.

I kept this letter as newsy as possible, apologizing for going two weeks without writing. But after I got back into the barracks later that day, I wrote again, taking her to task. And to make certain she understood how upset I was, I added: "I was about to be promoted, but the company commander said that because of your letter, I'm not going to be."

That wasn't true. But doggone it, she needed to have some idea of just how egregious it was for her to write such a letter.

From time to time over the years, I have thought of Captain Poppell. He may have lacked a formal education, but he enjoyed the respect of his men, and he really did put us first. When I think of the "Brown Shoe Army," he is the first person to come to mind. If he is still alive, I hope he has enjoyed his well-earned retirement. If he has gone on to Fiddler's Green, then I've no doubt that he's looking after his men there, as well.

The army was blessed with such men.

Bear Fight

When I was a low-ranking enlisted man at Fort Rucker, I went with some other soldiers to Dothan, Alabama, to get a tattoo. Dave got a heart with his girlfriend's name. Logan got an anchor. We were in the army and he actually wanted an eagle, but the anchor was cheaper. I watched both of them with some trepidation as they reacted to the pain of having it done.

My mother had asked me to never get a tattoo, as her brother had gotten one and she remembered how upset her mother had been. So, in order to placate my mother, I was going to have my tattoo on the bottom of my heel so neither she nor anyone else would ever know that I had one. I chose a pig's foot, because it was the cheapest, at $3.50, and I thought that a foot on a foot would be appropriate.

While I was awaiting my turn, another friend dropped by to see how we were doing, and he told us that there was a bear at the Farm Center—sort of a civic center in Dothan—and that for $2 you could wrestle him. If you managed to stay in the ring with him for three minutes, you would make $5.

I decided to eschew the pig's foot tattoo so I could wrestle the bear.

We went to the Farm Center, and I had a look at the beast. He wasn't all that big, but the trainer said he weighed three hundred pounds, which was only a hundred more than I weighed. Also, he had on leather gloves and a leather muzzle, so he couldn't claw or bite. And, he was tethered to one of the ring posts, so he could only come halfway across the ring. What could possibly be the problem?

I paid my $2 and climbed into the ring with him. I moved up and put my hand on his shoulder . . . *and he slugged me!*

My ears started ringing and I saw white flashes. Nobody said he was a boxing bear! He was supposed to be a wrestling bear!

I danced around for three minutes, making pretend moves toward him, but never getting close enough for him to hit me again or wrap his arms around me. When the bell rang, I gave a huge grin and went to collect my $5.

"You were disqualified, boy. You didn't maintain sufficient contact with the bear."

"All right, how about my two dollars back?"

"Sorry, boy, you signed the contract."

"That was just that I was getting in the ring of my own volition."

"It was a contract."

When the day was over, we went back to the barracks, where Dave and Logan showed off their tattoos and I became the butt of their jokes.

"You should'a seen ole' Vaughan stagger when that bear hit him. Funniest damn thing I ever saw!"

Well . . . I didn't want a pig's foot anyway

Plymouth: The Car, Not the Rock

Three of us went together to buy a 1947 Plymouth that looked really clean. The car salesman wanted $250, but we talked him down $10 so it would be an even $80 from each of us.

In those days it was easy to get a license plate in Alabama. You paid $3.50 and you got the plate, no questions asked, not even whether you owned a car.

We proudly parked our Plymouth in front of the barracks so all our friends could see. We carefully allocated the time we each had with the car, and I got to take it into Ozark twice.

The engine started knocking, and when we put it on a rack and dropped the pan, we discovered that either the car lot or the previous owner had put leather shims under the bearings. That quieted the car until the leather was used up, but the bearing and an out-of-round crankshaft meant the car needed a new engine.

That we couldn't afford.

We tried to sell it to a few people for as low as $100, but

by now, everyone knew about the car. As a last resort we took the car to a junkyard.

"A hunnert dollars," the man said.

We smiled broadly. At least we were getting some money back.

"I tell you what," I said magnanimously. "You can have it for ninety dollars, then you can give each of us thirty dollars."

"No, you don't understand. I won't be givin' you a hunnert dollars. That's what you're gonna have to pay me for me to take it off your hands."

We passed on the deal. I don't remember who came up with the idea of what to do with the car, but we agreed it was brilliant.

We went to the Fan Drive-In movie lot, took the license plate off the car, then after the movie concluded, returned to the barracks with some friends. As far as I know, that old Plymouth is still sitting there.

A Chance Meeting

In spring 1957, I went into town to pick up my date. She was still a senior in high school and was at a school club meeting. I was invited to come in and wait for her. Seated on the sofa watching TV was a man who told me his daughter was with Grace, so we watched TV together.

There was a WWII documentary playing, and it showed a tank firing. "Sherman tank," I said authoritatively. I wasn't in armor but I *was* in the army and needed to let this civilian know.

"That's a Grant tank," the man said.

"Sherman."

"Grant," this upstart civilian insisted.

"I'm in the army, and I've never even heard of a Grant tank," I said. "I'm telling you, that's a Sherman."

"The Grant was an earlier model: the M-3. The Sherman is an M-4." He smiled. "And I'm in the army, too." He was considerably older than me, so I assumed he was a sergeant.

"Oh. Well, Sarge, I apologize for arguing with you. I guess you know what you're talking about. Were you in tanks during the war?"

"Yes. I commanded an M-3 Tank battalion."

"You commanded a tank battalion?" I asked in surprise.

"I'm General Cairns. You're here to pick up Grace?"

"Yes, sir."

"My daughter Patty is a good friend of hers. What's your name?"

"Dick Vau . . . uh . . . PFC Vaughan."

Thankfully the front door opened then and Grace and Patty came in.

"Oh, Dick, I'm sorry you had to wait for me," Grace said. "Did you have to wait long?"

General Cairns chuckled. "Just long enough to learn the difference between a Grant and Sherman tank, right Dick?"

Wow. A general called me by my first name.

"Yes, sir," I replied sheepishly.

I got to know Patty pretty well after that, seeing her often with Grace.

Then, in December 1958, General Cairns was killed when his H-13 helicopter developed carburetor ice and crashed almost immediately after takeoff. I had not run into him again after that single encounter, but I did know Patty, and I felt a personal loss and a great sadness.

As an addendum, not too long ago, a writing friend came down to the beach and invited me to have lunch with him at one of the local golf clubs. When I arrived, he introduced me to Mike, his golfing partner for the day. Mike was a former jet fighter pilot who now flies for FedEx.

We talked flying for a while, and knowing that my aviation background was army, he asked me if I knew Cairns Army Airfield at Fort Rucker.

"Yes, I know it. I remember when it was Ozark Army Airfield." (I didn't tell him, but should have, that Ozark Army Airfield was where the movie *12 O'Clock High* was filmed.) "It was named after General Cairns because he was the CG of Fort Rucker when he was killed. I remember that vividly."

"My mother is General Cairns' daughter," he said.
"Patty is your mother?"
Mike's eyes grew large. "You know my mother?"
"I knew her," I said, a flood of memories rushing back to me.

I'm happy to say that I have subsequently heard from Patty . . . a brief touch with my past.

Duty Roster

Young soldiers today don't pull KP. So, for those who never had the rewarding episode of pulling KP, let me share the experience with you.

In the old army there was a duty roster that was posted on the company bulletin board. You were required to check the bulletin board at least three times per day. There you would find such things as a notice of a Saturday Inspection, a notice of an upcoming retirement parade, a posting of the latest promotions, and on payday the pay roster would be posted so you could see how much money you would be drawing that month. No privacy here ... you could also see how much everyone else in the company was drawing.

Here too, were the duty rosters: Post Duty Sergeant, Sergeant of the Guard, for the senior NCOs; CQ for the junior NCOs and corporals; Guard Duty, Fire Guard, and KP for the E-4's and below.

There were always instructions on the KP duty roster. Those selected for KP were required to report to the CQ the night before duty and tell them which barracks you were in and which bunk was yours. And you had to tie a white towel at the end of your bunk.

Generally, there were from 12 to 24 bunks in the bay, depending on whether or not they were double-stacked. After tying the white towel on the end of my bunk that night, I crawled into bed and tried, unsuccessfully, to tune out the card games, the loud talking, and the radio so I could go to sleep early. *Taps* would be played at 2200 hours, then lights would go out and it would grow quieter, but not entirely so. Finally, amid the snoring from a half-dozen of my barrack-mates, I drifted off to sleep.

"Vaughan? Vaughan? Get up, you've got KP." The voice and nudging were coming from the CQ runner.

"Go away, you've got the wrong bunk. I'm not Vaughan."

"Come on, Dick, get up. I've got four more to go."

Crap . . . it had to be someone who knew me.

I trudged through the night across the quadrangle to the mess hall. All the barracks were dark, and I was envious of the men who were still asleep in their bunks. The mess hall was the only building ablaze with lights, and when I stepped into it, I had to sign the roster. I was the first one there, so I signed up for DRO. There were three jobs on KP: DRO or Dining Room Orderly, Front Sink, which meant washing the serving trays (we ate from partitioned Bakelite trays rather than plates), and Pots and Pans. It was generally agreed that DRO was the best job, though the term *best* was relative. Pots and pans was the worst. DROs kept the salt and pepper shakers, napkin holders, and milk dispensers full. DROs swept and mopped the floor and bused the tables in the "top two grade" area, which was separated from the main dining room by a waist-high wall. In the days when I pulled KP, top two meant E-6 and E-7. The army had not yet added E-8s and E-9s.

The mess hall was permeated by the aroma of coffee. As the KPs arrived, and before the real work began, we would sit at the first table in the dining room to talk, drink coffee, and smoke cigarettes. I was not coffee drinker at the time and didn't smoke, so I would just sit there talking with the others.

"Vaughan, you're not smoking or drinking coffee, how 'bout coming back into the kitchen to clean up the floor where we just spilled some grease?"

The cooks were supreme. You could be an E-4 specialist and the cook could be an E-2 private, but when you were on KP the cook was absolute boss, regardless of the relative rank.

Finally the day would start. "Vaughan, start breaking eggs into this big bowl," the head cook ordered, pointing to a large aluminum bowl.

I was breaking them into a smaller bowl. If I came across a bad egg, as I would about every 20 eggs, I'd transfer it to another bowl to throw out later.

"Vaughan, what are you doing, breaking eggs into that little bowl?" the head cook asked. "I told you to break them into the big bowl."

"I'm culling out the bad eggs."

The cook picked up my cull bowl, which had at least four bad eggs, and tossed them in with the others. "I've been cooking in the army for 18 years . . . I ain't never saw no bad egg," he said.

Once while wiping down the scullery counter, the filthy sponge popped out of my hand into the big container we had just filled with orange juice. It floated around on top of the orange juice, leaving a wake of food particles.

"Sarge, the sponge just fell into the orange juice."

The cook came over, reached his hand down into the orange juice, and picked up the sponge. One more dip of his hand took out most of the food particles. He gave me the sponge. "Get the orange juice out on the serving line, and this won't happen again."

There was a period of time during the 1950s when it was decided to remove all the labels of canned and bottled products so as not to show brand preference. Once a cook was mixing a cake, and he was pouring in vanilla extract. He had put in half the bottle, then set it on the center island as he began beating the batter bowl. I had just brought him the beater, and as I looked at the bottle of vanilla extract, something about it didn't look right. I picked it up, put a drop on my finger, and tasted it.

"Sarge, this isn't vanilla extract. This is Worcestershire sauce."

The cook shrugged and emptied the rest of it into the cake mix. When the trays came back in through the scullery window after that meal, I noticed a lot of the cake was uneaten.

KPs worked constantly, with no letdown between meals. There were floors to mop, supplies to be unloaded from the delivery truck and placed in the pantry, grease traps to be cleaned (a perfectly horrible job), stoves to clean, and "edible" and "inedible" garbage to sort—yeah, I know, I never got used to the concept of edible garbage either.

Finally, a day that had started at 0300 that morning ended about 2130 that night. After a long, hot shower, I collapsed into the bunk, just as the first plaintive notes of *Taps* wafted through the company area.

> *Day is done, gone the sun,*
> *From the lake, from the hills, from the sky;*
> *All is well, safely rest, God is nigh.*

And to think that soldiers today will never have such memories.

Ships That Cross in the Night

I very much enjoyed my time as an instructor in the Army Aviation Maintenance Course at Fort Rucker. There was no field duty, so there would be no time spent out in the boonies, fighting rain and eating C-rations. It was pretty much an 0800 to 1700 job, then you were on your own.

Occasionally we would have additional duties, and I actually requested one of them. I learned that a West Point class would be coming to Fort Rucker for orientation. The reason I requested that duty was because a high-school friend of mine—in fact, we briefly had been debating partners—was a member of that class. My assignment was to be his personal liaison, and I was very much looking forward to it. But when I went to meet him, I learned he had left West Point shortly before.

"Would you take another Cadet?" I was asked.

"Yes, of course."

I was given Cadet Stillwell, a nice, friendly, eager young man who was eager to learn as much as he could about the Army Aviation School.

"Stillwell? There was a famous World War Two general named Stillwell. Vinegar Joe, I think it was," I said to him.

"Yes, he was my grandfather. My father is also a general."

"Wow, you have quite a tradition to live up to, don't you?"

"Let me ask you something, Specialist Vaughan. Were you drafted?"

"No, I enlisted." I laughed. "I've never even had a draft card. I came into the army before I was old enough to have one."

"If you enlisted, you came in by choice. If you think about it, with a grandfather and a father as career army officers, you might say I was drafted. From the day I was born, it was preordained that I'd be going to West Point."

"Do you regret it?"

Stillwell smiled. "No. Somewhere along the line I realized it wasn't as much an obligation as a privilege. I'm happy keeping up the family line."

The West Point cadets were there for a week, and I took Stillwell and a few others to my house one night for a barbecue. It was an enjoyable week.

Six months later I had one of the "coincidental encounters" that happen from time to time. General Joseph Stillwell Jr. was attending flight school at Fort Rucker, and because he was a general, his class consisted of one person. As it so happened, one of the classes I taught in the maintenance course was Aircraft Hydraulics, and that was also one of the classes the flight students took, so both the warrant officer candidates and the commissioned officers were frequently my students.

"Vaughan, look extra sharp tomorrow, you're going to have a general in your class," Sergeant Martell told me.

The next day, when I was introduced to the general as his instructor for the three-hour class, he returned my salute then asked, "Are you the same Vaughan who was liaison for my son a few months ago?"

"Yes, sir!" I replied, shocked, and to be honest, somewhat

thrilled that he had made the connection.

"I just spoke to him yesterday, and he said if I ran into you to say hello."

The class went well, and I felt as if I had just touched history, considering the fame of the general's father.

I wish I could end this story on a happy note, but I can't. Not too long afterward, General Joseph W. Stillwell Jr. was given command of the army in Vietnam. With a friend acting as navigator, the general decided to fly himself, in a vintage C-47 (military version of the DC-3) from San Francisco to Hawaii.

He never made it, and no trace of his plane was ever found.

An Encounter with a Lady of the Evening

David had the car, a 1951 Ford coupe. I rode shotgun, and Logan stretched out in the back seat. We were students in the helicopter crew chief's class at Fort Rucker, and we had our first three-day pass. As it so happened, it was also spring, and the Alabama schools had a week off. It was spring break—though this was 1956, and the term *spring break* had not even been invented. It was called AEA, or Alabama Education Association, meaning that the teachers would all be attending conferences and meetings.

David, Logan, and I were headed for the beach 90 miles away at Panama City, Florida, and Logan assured us, "There will be about three or four thousand girls there. We'll pick up three."

"What do you mean we'll pick up three girls?" I asked. "How do you just pick up three girls if you don't know them?"

Logan laughed. "You'd better let me do it. You just watch the master."

When we got to Panama City, we went to one of the beach hangouts. Logan and David both got a beer, while

I got a Coke, primarily because I was too young to drink. I didn't want to get into any trouble with the law, which would have resulted in an Article 15 from the army, and I sure as hell didn't want that. But to be totally honest, I didn't really like beer.

"There are three girls over there by the pinball machine," Logan said. "This is just perfect."

"What's perfect?"

"You just wait until I start playing. The lights will flash, the bells will ring, and they'll come running. Girls are impressed by someone who really knows how to work the machine."

David and I watched. Logan tried three times, but the machine tilted all three times. Then, slapping his hands on it, he came back to the table.

"The machine is fixed," he said. "They don't want anyone to win."

"What about the girls?" David asked.

"What about 'em?"

"Did you ask if they wanted to come over?"

"Nah, I checked 'em out." Logan shook his head. "They aren't anyone we'd be interested in."

We tried three or four more places, but Logan struck out every time. That night we planned to sleep on the beach to save money, but at about two in the morning I was awakened by a kick in the side. When I opened my eyes, all I saw was a bright flashlight shining in my face.

"Get off the beach!" a gruff voice said.

"Why?"

"Because if you don't, you're going to jail."

There were two policemen standing over us. I had no idea that you couldn't sleep on the beach.

We spent the rest of the night in David's '51 Ford. Since it was David's car, he got the back seat. I tried to sleep in the driver's seat, but this was long before they reclined.

By midafternoon the next day, and after another half-day

of failure in picking up any girls, Logan announced that he knew a surefire way.

"You mean like the pinball machine?" I said smugly.

"No. I mean like going to a cab stand."

"A cab stand? Why do we want to go to a cab stand?" David asked. "I have a car, remember?"

"Yes, but a cab driver will know where to go."

"Cab drivers know where to find girls?" I asked.

Logan smiled. "Yeah. The kind of girl you pay for."

"You mean a whore?" David asked with a gasp.

"Yeah, why not? We are three soldiers on a three-day pass. That's what soldiers do on a three-day pass."

"I don't know," David said. "I've never had to pay for it before."

"I've never had to pay for it either," I said. That was true. I never paid for it because, in fact, I had never done it before.

Against my better judgement, and because I didn't know what to do while they were gone, I agreed to "pay for it."

The cab driver took us to a house trailer out in a growth of palmetto bushes. He went inside for a moment, then came back out.

"Okay, boys, she's all ready for you," he said with a knowing smile.

Logan was first, and while he was in the trailer, I sat in the back of the cab looking out at the waving palmetto fronds. All I could think about were the movies the army showed on hygiene, all of which emphasized the danger of sexually transmitted diseases. I was terrified! What if I got something?

Logan came out, tucking in his shirttail. "Boys you're in for a treat!"

"I'm next," David said quickly, and I didn't argue with him.

For the next 15 minutes or so, Logan entertained the cabbie, and I suppose he thought me, with stories of all the women he had been with. "But this one ranks right up there near the top," he said.

David returned then, and it was my time. I wanted to say that I wasn't feeling good, that I had a headache, anything to keep me from having to go into that trailer. But I couldn't come up with an excuse, so I reluctantly went inside.

To my surprise, there was a man inside, sitting in a rocking chair and reading a newspaper.

"Uh?" I said.

"She's in there," he said with a jerk of his thumb, not taking his gaze from the paper.

The chair was squeaking as he rocked, and I stepped up to the door.

"Hello?"

"Come in, honey, come in."

The woman was sitting on the edge of the bed, smoking a cigarette. She looked to be in her late 30s or early 40s, which to me, at the time, was a senior citizen. I could still hear the rocking chair: *Squeak, squeak.*

"Are you the last one?"

"Yes, ma'am."

She ground the cigarette out in a fruit jar lid on the bedside table. "Okay," she said.

I just stood there.

Squeak squeak.

"Well?" she asked.

"I, uh, don't know what to do."

"Good Lord, don't tell me you're a virgin."

"No. Only girls can be virgins," I responded knowledgeably.

"Honey, have you ever done this before?"

"No, ma'am."

"How old are you?"

"Seventeen."

"Babies," she said. "They are taking babies in the army now."

"I, uh, don't want to do this."

"All right."

"But I don't want my friends to know that I didn't do it, so I would like to stay here about 15 minutes."

"Honey, if you stay here, whether you do it or not, it's going to cost you five dollars."

"Okay."

"Do you want a cigarette?"

"I don't smoke."

"Why am I not surprised?"

I sat in a chair, and she sat on the edge of the bed, smoking another cigarette. She asked me a couple of questions, such as where I was from, but I wasn't in much of a conversational mood. Finally she ground the second cigarette out beside at least four others.

"Time's up, honey."

I left the room. The man was still rocking.

"She knows what it's all about, don't she, sonny?"

"Yes, sir," I said. "She sure does."

On the way back to Fort Rucker, I had the back seat while David and Logan talked about what a wonderful weekend we had. I pretended to sleep.

Sling Loading the H-19 Helicopter

One of the classes we taught was a crew-chief class for the H-19 helicopter. This was only three years after the Korean War, where the H-19 had proved the concept of using helicopters to move troops and equipment. At the time it was one of the biggest helicopters the army had, and while the sling-load weight was nothing compared to what today's helicopters can carry, it functioned fairly well.

One of the classes we taught was how to hook up the sling load, both to the crew on the ground and the crew chief who rode back in "the box." It was a fairly involved procedure that required good hand-eye coordination, as well as coordination between the pilot and the sling-load crew.

We practiced at a place called Tac Able, which was a concrete area about the size of two football fields sandwiched between barracks and classrooms on the southeast corner of the post. I was in the box with the student crew chief, and other instructors were on the ground with the rigging crew. Our load was an old jeep—sans engine, transmission, and everything that might add weight.

We had been picking up the jeep, making a wide circle, coming back to drop the load, then landing to pick up a new student. It had been going well all day, when we picked up a new student. The pilot took off, made a wide swing, then came back to pick up the jeep.

I was sitting in the open door, watching the student crew chief as he lowered the cable and made the connection. It went without incident.

"Sir, the load is secure," I told the pilot, a fellow named Mr. Kirby, who was up in the cockpit above the box, so we had no direct view of each other.

The pilot took off, and I went through the rest of the procedure with the student.

"The pilot has an emergency drop, in case the load gets unstable," I said. "But if for some reason his emergency drop doesn't work, you can cut the load manually by pushing on this lever here."

We were at about 500 feet when I gave that little spiel, which I had done so many times by now that I did it by rote.

"This one?" the student asked—*as he pushed the lever!*

With a sickening sensation, I watched the jeep break away from the sling and start plummeting 500 feet. I leaned out of the door as the jeep got smaller and smaller, giving a prayer of thanks that we were still over Tac Able with nothing below us but concrete.

"Vaughan! What the hell just happened?" Kirby asked.

"Sir, we just dropped the jeep."

"Did the cable separate?"

"No, sir. We . . . uh . . . jettisoned it."

"Get out of my helicopter!"

"Yes, sir, as soon as—"

"Now, dammit! Get out of my helicopter now!"

"Sir, we're five hundred feet in the air."

"Jump!"

Mr. Kirby was very upset, and I was getting frightened. "Sir, we can't . . ."

The pilot lowered the nose and flew quickly away from the area, then set down.

"Get out before I come back there and throw you out!"

"We're at least two miles from the site," I explained.

"Now!" he yelled.

The student and I got out—it seemed the thing to do—then we watched as the helicopter took off, leaving us behind.

"Boy, he was mad, wasn't he?" that doofus student said.

"Yeah, aren't we?" I replied. But I don't think he caught the subtlety.

Wound Stripes

The last World War I veteran to serve on active duty in the U.S. Army was Master Sergeant John Wooley, and I served with him at Fort Rucker in 1956. Being around him was like reaching out to touch history . . . and he was treated like the icon he was.

In those days when your car was registered on post, you were issued a bumper sticker: blue for officers, red for enlisted, and green for civilians. Blue number one was the commanding general of the post. Red number one was supposed to be for the sergeant major, but at Fort Rucker it was reserved for Sergeant Wooley.

Wooley was director of the post museum, but clearly he was the most interesting part of the museum.

We were in the Headquarters, Staff, and Faculty company, a service company that had no TO&E (table of organization and equipment) structure. That is to say, we had a company commander, executive officer, and first sergeant, but no platoon leaders or platoon sergeants. Actually we didn't even have platoons, but on those once-a-month occasions when we would have a Saturday morning formation, we fell into platoons by rank: E-1 through E-4 in one platoon, then separate platoons for E-5, E-6, and E-7.

The senior-most person within each platoon would post to the front of the platoon for the formation. One Saturday morning, Wooley moved to the front, and a recently arrived E-7 called out, "Hold on there, Sergeant. You didn't check date of rank. Mine is June 12, 1943. What's yours?"

Wooley didn't even turn around. "October 14, 1918."

"Never mind," the E-7 said as the rest of the company laughed.

Generally we would stand these formations in our class A uniforms, with "all ribbons." The only ones who had ribbons were those who had actually been in combat in Korea, which was only three years earlier, or World War II, which was 11 years earlier. And of course Master Sergeant Wooley, with ribbons from three wars.

Wooley also had a small upside-down chevron on his left sleeve cuff. We had a brand-new second lieutenant who, certain it was an unauthorized adornment, moved over to address him.

"Sergeant what is that on your left sleeve cuff?"

Master Sergeant Wooley, who was standing at attention, continued to stare straight ahead.

"Wound stripes, Sonny. WWI wound stripes."

You could hear the company's laughter all over the post.

Wedding Bells, Without the Bells

While stationed in Alabama, I went to all the football games at Ozark High School (now Carroll High). One reason is because I love football. Another reason was because the Ozark High football uniforms looked exactly like the Sikeston High uniforms, so it was almost as if seeing games in my hometown.

But there was still another reason. The head twirler was a very pretty redhead, and I liked the way she tossed her hair around as the band was marching. Once, when they came right up to the bleachers where I was sitting, she blew back a bit of hair that had fallen forward, and I pretty much fell in love that very moment.

I managed to get a date with her, and we went to a few movies, including a couple at the local drive-in. She was the daughter of the chief of police, and she was very popular and self-assured. I enjoyed being with her. This was the girl named Grace whom I had been waiting for when I met General Cairns.

Then I learned that I had been approved for an appointment to West Point. Excited, I drove into Ozark, not only to share the news but to take her to a movie. We had been dating for about three weeks, and I was comfortable going into her house and interacting with her parents.

When I arrived to pick her up, her father, Tom, was in the living room. I greeted him jovially, but his reply was so short that I knew he didn't want to extend the conversation.

"Are you ready?" I asked Grace.

"Where are you going?" Tom asked her.

"I'm going to a movie with Dick."

"No, you aren't. The governor is in town, and you are going to the governor's dinner."

Grace and her father got into a heated argument over whether or not she was going out, and it reached the point where Grace called him a dirty bastard and he slapped her. She went careening across the floor, hit the wall, then slid down. It was pretty obvious that she was overacting, but it made me very uncomfortable to be there.

"Dick!" she cried out. "Are you going to let him hit me like that? He's always beating on me, and I've no one to protect me. Are you going to let him hit me again?"

"Hell, no, I'm not," I called in my most macho voice, and I hit her father with a roundhouse right that knocked him onto the TV stand, breaking the TV. And remember, this was the chief of police.

"You son of a bitch!" he said. "Wait until I get my gun."

He started toward the back of the house, and I ran out the front door. As it happened, I knew the people who lived next door, and I started pounding on their door, which fortunately was opened quickly.

"Harold, I need a gun! Have you got a gun? Tom is going to kill me, I need a gun!"

Harold invited me in to calm me down. A moment later, Grace came in and sat on the sofa beside me.

"Don't you worry," Grace said. "I'm not going to let Daddy shoot you."

Then Grace's mother came in. "Tom is outside, and he wants to see you."

"What? I don't want to see him!"

"Don't worry, he doesn't have a gun, and he wants to apologize to you."

Reluctantly, but with Grace and her mother to protect me, I forced my bravest expression and confronted the chief of police.

"I'm sorry I said I was going to get my gun," he said.

"I'm sorry I hit you."

"Grace, if you want to go out with this boy, go ahead. You don't have to have dinner with the governor."

"That's not good enough," Grace said.

"What do you mean?"

"We're going to get married."

I sort of went numb. That was not a subject we had ever discussed. She was still in high school.

"That right, boy? You going to marry my daughter?"

He had just threatened to get a gun, remember. "Uh, yes," I said. I knew it would take three days, and this would pass over, especially after I told her about my appointment to West Point.

"We're going to get married tonight," Grace said, as if reading my mind. "We'll go to Mississippi. There's no waiting period there."

"No daughter of mine is going to get married in Mississippi," Tom declared.

"Then you better call Governor Folsom and have him arrange it."

As Tom headed into their house, I remained unconcerned. I knew the governor wouldn't agree, and tomorrow this would all blow over.

A few minutes later, Tom came back out. "Okay, the governor called the health office up in Troy. Go up there and get your blood tests; they'll work them up tonight." He handed me a slip of paper with the address of the health office and the name and address of the judge we were to see afterward.

Wait a minute, I thought. Now this really was getting out of hand.

"Come on, Dick, let's go," Grace said.

During the 30-mile drive to Troy, I wondered if I should mention West Point. She talked nonstop about how she would be the first in her class to get married, and when we went to the senior prom, it wouldn't be a date—we would be man and wife.

It was getting pretty late by now, and I was sure the health office wouldn't be open, and that would be what I needed to slow this down.

My jaw dropped as we drove up to the building and found the county health official waiting in his car.

"Are you the ones that the governor called me about?"

"Yes," Grace said excitedly.

We went in to have our blood tested, and as Grace continued to talk, I continued to sit there like a coward, unable to tell her I didn't want to do this.

After a while the health official came back with a piece of paper. "Show this to the judge when you get the license," he said.

"You mean it's all right, we can get married? We have the same kind of blood?" Grace asked.

The health official looked at her and muttered, "Uh, yeah."

It was after 10 o'clock by the time we reached the judge's house. At first, I thought he must be in bed, but when we walked up the steps, the front door opened.

"Your dad called me," the man said. "Do you have the paperwork from the health office?"

"Yes," I said.

We followed him into the house and over to the dining room table, which was covered with papers. He ruffled through them, then pulled out a license certificate.

"Grace, honey, how old are you?" he said as he started to fill out the form.

"I'm 17."

"No, honey, you're 18."

"How old are you, boy?"

"18," I said. At least I was old enough.

"No, you're 21."

He gave me the license. "There are no names here," I said.

"Reverend Snellgrove will put the names in when he marries you." He smiled at Grace. "He's over at your house now."

When we got back to Grace's house, there were cars in the driveway and parked out on the street. When we went inside, the house was full of people, which I thought was very odd, because it was nearly midnight.

Everybody greeted Grace effusively.

"Preacher, you ready to get these kids married?" Tom asked.

Kids. He said kids. At least he realizes that.

"Soon as I finish this cup of coffee, Tom."

I looked into the cup. It was half full. Only half a cup between now and being married.

Finishing his coffee, the preacher stepped into the wide doorway between the living room and the dining room. With a little wave, he invited us up.

"Dearly beloved, we are gathered here to join this woman, Grace, and this man . . ." He hesitated. "What's your name, boy?"

"Robert."

"Robert?" Grace said looking at me. "I thought your name was Dick."

"My name is Robert, but I go by Dick."

Grace turned to address the 20 or so guests. "Now, that's the strangest thing, isn't it? His name is Robert, but I've only known him as Dick."

The preacher finished, then pronounced us man and wife. The kiss was the first one we had during this entire night.

Tom came up to me then. "Okay, boy, you've got her.

What are you going to do with her?"

"I . . . I don't know, I haven't thought about that. I can't take her to the barracks."

"How much money have you got?"

"Eleven dollars."

"I don't mean how much on you. I mean, how much money do you have?"

"Eleven dollars," I repeated.

He left me standing there, alone. Nearby, Grace was surrounded by people. When he returned, he said, "You can stay here tonight, I just made arrangements for you to move into the cabin behind Holman's house."

The next morning, to my shock, I got a telephone call from my mother.

"Dickey, I just got a message that said I should call you at this number, that you had news for me. Please don't tell me that you have gotten engaged. You're much too young for that."

"I'm not engaged, Mother, I'm married."

Back at Ya

The H-19 and H-34 were the premiere cargo helicopters in the army throughout the 1950s. These aircraft were susceptible to something called ground resonance—violent oscillations that build very quickly to catastrophic failure, which can result in total destruction. It can only happen in helicopters with three or more non-rigid blades.

When I was an instructor in the crew chief course at Fort Rucker in the fifties, this was something we had to teach to the mechanics, because many crew chiefs would start the helicopter on the ground for routine checks and to track the blades. They had to be aware of the condition, so I was tasked with writing the lesson plan for this class.

I began researching the phenomenon, but at the time there was very little in print about it. I was growing increasingly frustrated, until my friend Gilbert came up with an idea. Gilbert had just bought a set of the *Encyclopedia Britannica*. One of the sales pitches they were using then was a free research program. You could send them a question and they would research it, then send you a complete analysis.

That, I thought, was a great idea, so I had him write to the encyclopedia company. "Please research and send me the results of everything you have on ground resonance in helicopters." Then, I waited for the answer.

Two weeks later, Colonel Knox called me into his office.

"Specialist Vaughan, how are you coming on your lesson plan for ground resonance?"

"I'm still working on it, sir."

"Are you finding all the information you need?"

"Yes, sir," I said, anticipating the response from my request for more information.

"We just got this letter from Encyclopedia Britannica," the colonel said, sliding it across his desk.

"Great!" I said. This was exactly what I was waiting for.

"It's very flattering that they would come to us, don't you think?" Colonel Knox asked. "Don't let the army down, now."

"Sir?" I said, somewhat confused. I took the letter and read it.

"We have recently received a request to research Ground Resonance in helicopters. As often as we can, we like to go to the actual source for our research, and as Fort Rucker, Alabama, is the home of army aviation, we are certain that you will be able to provide us with the expert information we need."

"Write a response to this," Colonel Knox said, smiling broadly. "Damn, I think it's neat that they came to us. I'll send it back on official letterhead stationary . . . with our motto: Above the Best."

I wrote what I had been able to gather so far, gave it to Colonel Knox, then finished my lesson plan, knowing that I wouldn't be getting any new information.

About two weeks later, Gilbert brought me his response from Britannica and started to read it to me.

"Wait, listen to this first," I said. "Ground resonance can be induced by a shock to the aircraft, such as a weak oleo strut that might initiate the first wave of airframe vibration. This is exacerbated by a disproportionate movement of the rotor blades in their plane of rotation. A failure in the damping system will generate unbalanced lift, thus shifting the center of gravity on the axis of rotation, with the possibility of total airframe destruction."

"Wow, that's word for word. How did you get this letter before I did?" Gilbert asked.

"Because I'm the one who wrote it."

The article appeared verbatim in the next issue of the *Encyclopedia Britannica*.

My First Book

I was an E-4, an instructor in the aircraft maintenance course at Fort Rucker in Alabama. General Cairns, the post commandant, issued an order that every permanently assigned officer must write an article for the *Army Aviation Digest*. My OIC (officer-in-charge) knew that I liked to write, so he gave me $25 to write the article for him. I did, and it was published and won an additional $25 for being the best article of the month. To his credit, my OIC also gave me that money.

Soon I was writing articles for several other officers. It was the beginning of what would become a lifelong "ghost writing" career. I won best article of the month three more times and best of the year, which was $100. Soon I was making more money writing articles than I was earning from my army pay.

It was then that I met William E. Butterworth, who at the time was chief of the publications division at Fort Rucker and now is known to millions of readers as best-selling author W.E.B. Griffin. I told Bill that I wanted to be a writer and that I was currently working on a Civil War novel.

"We don't need another Civil War novel," Bill said. "Write something like this." He tossed a mass-market paperback novel across the desk to me.

I was very young and very full of myself. I didn't want to write a paperback novel, I wanted to write a *real* book. But I followed his suggestion, wrote a paperback, and sent it off to an address I got from inside a book cover.

One month later, I got a letter back. "We would like to publish your book. Contract to follow."

Wow! I was 19 years old and soon would be rich and famous! Hemingway, Faulkner, Spillane had nothing on me! I bought a new car.

Then the contract arrived. I was to receive $100 in advance and royalties of one-half cent per copy sold.

The book wound up making a grand total of $180. I convinced a sergeant to take over payments on my new red-and-white Chevrolet Bel Air in order to avoid the embarrassment of having it repossessed.

My next book brought $150, and within 10 books, I was up to $600 per book. It was a slow start... but it was a start.

To Shoot, or Not to Shoot. That is the Question

Walking guard wasn't nearly as odious as pulling KP. Of course, walking guard on a frigid winter's night or in a pouring downpour wasn't pleasant, but if the weather was good and temperate, it really wasn't all that bad.

The makeup of guard consisted of privates of the guard, commander of the relief, sergeant of the guard, and officer of the day. At one time or another during my army career, I managed to fill all four slots.

The duty day started with guard mount. This required all the privates to stand inspection in class "A" uniform. Boots had to be shined, belt-buckles and brass polished, hair neatly cut and closely shaved. In addition, your weapon (the M-1 rifle during my time) had to be spotless. You stood mount in three ranks (three reliefs), and the inspecting officer passed down, pausing in front of each soldier. As he did so, you brought the weapon up to "inspection arms" Slamming open the bolt with your left hand and holding it at an angle across your body. Then you watched the officer's left shoulder. As soon as it moved, you jerked your arms down, sharply, by your side, leaving the rifle hanging in midair. Of course, it didn't actually stay there, because the officer's left hand would grab the forestock as soon as you let go. It became sort of a contest between the private and the inspecting officer so that their hands would never be on the rifle at the same time. Returning it was the same, you had to watch carefully because as soon as the rifle was in position, the officer would jerk his hand away, expecting you to keep the rifle from falling.

Then would come the questions:
"Who is the army chief of staff?"
"General Maxwell Taylor, sir!"
"What is the Fifth General Order?"
"To quit my post only when properly relieved, sir!"

There was always a supernumerary in the rank: one man more than was needed to walk all the posts. That was to allow the officer of the day to select the colonel's orderly, who spent the whole time in the guardhouse without having to walk his tour.

There were three reliefs, each relief walking two hours, then off for four, then walking two more hours. First relief went on at 1800 (six in the evening) and walked until 2000. Second relief walked from 2000 until 2200, then third relief from 2200 until 2400. At that time, first relief came back on duty, walking until 0200, then second relief until 0400, then third relief until 0600. If you had the last relief, you watched as the sun rose, you heard *Reveille*, and you could smell the breakfast bacon.

Walking your post could be pleasant—you were all alone in the middle of the night and could think things through. I was a platform instructor at Fort Rucker, so when I was walking in the middle of the night, I would practice teaching my class . . . aloud. It helped to pass the time, and I'm happy to say that I was never caught doing it, which would have been embarrassing.

Sometimes I would walk guard around the officers' club,

dreaming of the day I would be one of those men going in and of the club. The day finally came when I became an officer, and I remember the thrill I experienced the first time I went into the club. I'm sorry to say that officers' clubs no longer exist.

My favorite post at Fort Rucker was post number five. It was the ammo dump, at least three miles away from the main post and surrounded by a very high chain-link fence topped with concertina wire. This was a serious post, and we were given a loaded shotgun. There also was a telephone so you could call into the guardhouse if you saw something suspicious. Even better, it was a class A phone—one with an outside line—which meant I could call my Grace if I had one of the earlier shifts.

There was only one road out to post number five, so neither the officer of the day, the sergeant of the guard, or the relief change could sneak up on you. Also, you could tell by the sound whether it was the OD's jeep, or the relief change three-quarter-ton truck, which had a universal gear that made a very distinct whine.

That meant you could sit down, lean back against one of the ammo bunkers, and take it easy. One night as I was doing this—to be honest, I think I may have dozed off—I felt something rubbing against my boot. When I opened my eyes, I saw it was a skunk! I was terrified it was going to spray me, and I raised my shotgun to shoot it. Just before I pulled the trigger, however, I remembered that I had once hit a skunk with my car, and the odor stayed with the car for quite a while. Would this skunk spray me if I shot him? I knew at point-blank range he wouldn't be able to do it purposely, but would it be released as he died?

I sat there in total fear for at least 15 minutes, barely able to breathe. He sniffed my boots, my legs, my arm. Then he looked up at me, staring straight into my eyes. "What are are you goin' to do now, huh, big guy?" he seemed to ask. Then, with a shrug, he walked away . . . just as I heard the relief truck arriving.

The Great Gaffey

I went to Korea on board the USNS *Gaffey*, a troopship that had been transporting troops since 1942. It could carry 4,000, though I hasten to add that not all of them traveled in comfort. I was one of the uncomfortable ones, down in the bowels of the ship where we slept in racks stacked six high. I had no experience of going to sea, but I was sure something as big as the *Gaffey* would be as steady as a house.

I was wrong. The *Gaffey* started rolling in San Franciso Bay, and I was seasick by the time we passed under the Golden Gate Bridge.

There were several things about the ship, other than the rolling and pitching, that worsened my sickness. For one thing, going to the latrine, or the *head* as the sailor called it, was particularly hard. Instead of individual urinals, they were long, trough-like urinals, and as you stood there with the ship rolling, you could see the output of the bladders of men before you, rolling from one end of the urinal to the other.

Going to chow was difficult, too. From my compartment, we had to take a catwalk over the engine room, and the smell of hot diesel oil was enough to turn me around and give up any idea at all of chow.

I managed to subsist on oranges that friends would bring me. For some reason, oranges were not only the one thing I could safely eat, I actually began craving them.

Then one rare day when the sea was relatively still, I made it all the way to the mess hall. When we had first boarded the *Gaffey*, we had been given chow cards that were punched at each meal, to make sure we didn't double up. We had been at sea for eight days, and this was the first time I felt like eating anything beyond an orange. When I showed my card, the sailor looked at me in surprise.

"There are no punches in this card," he said, confused.

"I know. I just came aboard last night," I lied.

"Oh," he replied as he punched my card.

There were some wives on board, but we never got to see any of them. Apparently, there was a "queen of the voyage" election being held, or at least that's what Sergeant Flynn told us.

"Men, Susie is the wife of a Spec-5 in Japan, prettiest little thing you've ever seen, young and so sweet. I had to practically beg her to run for queen of the voyage. It costs a dollar a vote, and whoever raises the most dollars will be the queen. She's running against some lieutenant's wife. What do you say we collect enough money to get her elected?"

His suggestion raised at least $100 from our compartment alone, and we were number two of eight compartments in the bowels of the ship.

"This should do it, boys," Sergeant Flynn said with a big smile.

Two days later, he returned with an angry expression on his face.

"You won't believe this," he said. "There is some old biddy who is a colonel's wife, and she says that no pig wife of an enlisted man is going to be the queen, and she'll see to it. Boys she's raising money for this thing like nobody's business. I'm sorry now I talked that sweet little girl into entering the contest. I mean, I had no idea she would be insulted like this."

"What!" Someone shouted out. "She called an enlisted man's wife a pig?"

"I'm afraid she did boys, and she's raised more money than we have. I'm sorry I brought you into this."

"Here's ten dollars," someone said. "I know there are enough of us that no colonel's wife is going to just buy the crown for that lieutenant's wife."

"Here's five dollars," another said.

Everybody in the compartment put in at least another dollar.

Three days later, Sergeant Flynn came back, this time grinning broadly. "Boys we're almost there. I heard the colonel's wife say she couldn't raise no more money, and our girl is just a few dollars behind. I know we can get her over the line now."

Again the men riding in the eight compartments in the bowels of the ship came up with more money.

The next day Sergeant Flynn came back with a big smile. "Boys, we did it. Our little girl won. I tell you, it has really pissed off all the officers and their wives, too. And when Carol Lee, that's her name, leaves the ship in Japan, she's going to say thank you to all of you."

"I'm glad she won, bless her heart," someone said.

"Yeah, I'm glad we gave that old colonel's wife the finger."

A few others laughed.

When the ship docked in Japan, we went up on deck, looking for Sergeant Flynn, but we couldn't find him. The wives began leaving the ship then, and someone pointed to a pretty, very young woman.

"Hey, I bet that's her. Hey, Carol Lee, come talk to us!"

Several others shouted, as well, and frightened, she hurried off the ship.

One of our guys saw one of the crew of the *Gaffey*.

"Hey," he said. "How come the queen of the voyage didn't come talk to us?"

"What queen of the voyage?" the sailor asked.

"You know, the one we paid money to buy votes for."

The sailor laughed, and shook his head. "Boys, I think you've been taken. I've made this crossing ten times now, and there has never been any queen of the voyage."

We started counting up the money as best we could, and we figured that Sergeant Flynn, if that was even his name, pocketed about $8,000 on the phony "queen of the voyage" election.

A Streetcar in Tokyo

Those of us who had arrived on the USNS *Gaffey* were given the opportunity to visit downtown Tokyo. Afterward, I realized I didn't have enough money to take a taxi back to the ship, and I didn't trust myself to find my way by bus or trolley. So I stood on the corner asking passersby if they could speak English. Eventually a man in his forties carrying an umbrella said that he could. I opened my hand and showed him the Japanese coins I had.

"I have to get to my ship in Yokohama. Will this get me there?" I asked.

He said yes, then signaled me to follow him. He stepped out into the middle of the street, flagged down a trolley, and spoke to the motorman.

"He will help," the man said. He reached for some money and gave it to the motorman.

"Take some for yourself," I said, but he smiled and waved it off.

With the trolley stopped in the middle of the street, the motorman asked a woman in the front seat to move—which she did, all the while smiling at me—and he motioned for me to sit there. After a long ride, the motorman stopped, announced something to the passengers, then signaled for me to come with him—he had to signal, as he spoke no English. Leaving his trolley sitting there, he led me about two blocks, where he stopped another trolley, spoke to the other motorman, and gestured for me to get on board.

After another 45-minute ride, that motorman stopped, left his trolley loaded with passengers, and walked about a block, then pointed to a ship—the *Gaffey*—just a short walk away. This was only 14 years after the war ended, and I still have vivid, very positive memories of that incident.

Curriculum Vitae

My army career was in two parts. The first—the very early days in Fort Rucker, my time in Korea, and a tour of duty with the 4th Army Flight Detachment in Texas—were spent as an enlisted man. Then, as a warrant officer, I was at Fort Campbell on the Kentucky-Tennessee border, the Schweinfurt Army Base in Germany, and three tours in Vietnam, sandwiched between tours at Fort Eustis in Virginia.

In some ways, the differences between my two periods of service were like night and day. The unifying factor, of course, was me. I was the same person in both manifestations, and regardless of the stimuli I encountered, my reaction was pretty much the same.

I would like to tell you about the Korea I knew. First, understand that the Korea of the late fifties is absolutely nothing like the Korea of today. When I was there,, we couldn't eat off base because every civilian eating establishment was off-limits due to health reasons. We also couldn't drink the water.

The little village outside the base where I was stationed was Tong Du Chon. The entire village was squalid: no electricity, no water or indoor bathrooms, no paved roads or sidewalks, and the only motor-powered vehicles belonged to the U.S. or Korean armies. At least half of the houses were made from flattened beer cans, the other half from recycled boxes and crates. There was nowhere to go to escape the odor of fish and feces, and it was as if the nation's entire economy was based upon stolen cartons of Pall Mall cigarettes.

Compare that with Korea today. Seoul, with its towering skyscrapers, elegant shops, and broad avenues filled with automobiles built right there in Korea, is the equal of any city in the world. To think that they came so far in the time since I was there is certainly a testament to the resourcefulness, work ethic, and intelligence of the Korean citizen. I am awed by all they have done.

Even Tong Du Chon, the squalid little village, is a bustling metropolis today, with 15- to 20-story buildings and paved four-lane highways.

I want to share a little story. One morning before daylight, about six of us were going over to the flight line to prepare our aircraft for the day's operations. As we crossed over the MSR (main supply route, which is army speak for highway) and walked along a ditch that separated us from the barbed-wire boundary of the airfield, one of the men held up his hand.

"Whoa, hold it! I hear a kitten. Damn if somebody hasn't thrown away a kitten. I hate that." He started down into the ditch.

"Lingle, you ain't goin' to have no damn cat in the barracks now, you hear me?" one of the other men said. "You think we're goin' to want a cat sneakin' around, poopin' and peein' all over the place?"

"Don't worry about it. Cats mostly take care of themselves."

"Then let it be. He doesn't need you."

"Unless it's a kitten and needs . . . *Holy crap! It's not a cat!*"

"What is it?"

"A baby!"

That got everyone's attention, and we hurried down into the ditch to discover a box containing a baby who couldn't have been more than a couple of days old. There were some old rags in the box, but the little boy was completely naked. It was a good thing it wasn't cold, but there were plenty of mosquitoes, and the baby already had several bites.

We took the baby back to the orderly room, and the CQ called the company commander.

"I bet it's hungry."

"How are we going to feed it?"

We went from the orderly room to the mess hall, where one of the cooks warmed some milk. A clean T-shirt was dipped into the milk, then stuck into the baby's mouth. He sucked on it hungrily.

The company commander joined us in the mess hall and listened to our story. "Can I keep ' im, Cap'n," Lingle asked. "Nobody wants 'im, or they wouldna thrown 'im away."

"I don't think that's very practical," Captain Kirby said. "We'll take him to the orphanage."

The orphanage the CO was talking about was one that our company sponsored. Once a month, as we passed through the pay line, we dropped money into a box marked "For the Orphanage." We also provided them with food and blankets from our own supplies.

"We cannot take him," the orphanage director said when we got there.

"Why not?"

"This is not the way the children come to us. The government supplies the children."

"Tell the government they will supply this child . . . or they will also be responsible for supplying one hundred percent of your funding and your food and blankets," Captain Kirby said.

"Ha! Look at the ole man give it to 'im," one of the men said.

Captain Kirby's gentle persuasion worked, and the orphanage took the baby, who wound up with as many names as there were people who knew about him. Lingle called him Tom Sawyer, the only child's name in literature that he knew. A few called him Moses, because of the way he was found. I don't think I came up with a name for him then, but I have since referred to him as Mong-Koo Chung, who until recently was CEO of Hyundai Motors.

All right, the boy probably isn't Mong-Koo Chung, and I have no idea what actually happened to him in the 60-plus years since we found him. But isn't it nice to wonder what could be?

Crying in the Mashed Taters

I was in Camp Casey, Korea. It was about 2100, well past supper, and I had "the munchies," so I walked down to the NCO Club to buy a sack of French fries. I'm not talking little bags like you get at MacDonald's, I'm talking a pretty substantial bag, one as big as the companion bag, which held a can of Coke.

I knew that with this large a bag of French fries, I would have to share them with some of the others in the hooch (barracks), so I selfishly gave them a really good soaking in Tabasco sauce. I wasn't the only one who liked Tabasco, but this simple little trick eliminated more than half my barracks-mates.

As I started back to the company area, I heard the song "Ramblin' Man" by Hank Williams. There were no buildings nearby, so I walked over to see where the music was coming from. I saw an old master sergeant—he was old to me at the time—who was sitting cross-legged on a little hill, listening as the music played on a small, battery-powered record player. On one side of him was a stack of 45 rpm records, and on the other side were several empty bottles of OB, a Korean beer favored by the GIs. As "Ramblin' Man" finished, he took off that record and put on "Hey Good Lookin'," then "I Saw the Light." I realized that every record in that pile was by Hank Williams.

"My dad really likes Hank Williams," I said.

The master sergeant took the final swig of another bottle of OB, then tossed it aside. He looked at me with eyes that wouldn't have been more intimidating if he had been staring at me over the barrel of a Model 1911 .45 pistol.

"Boy, do you like Hank Williams?" he asked. I never heard another deep voice like his until the first movie I ever saw with Sam Elliot.

"Yes, Sergeant, I like Hank Williams," I answered, knowing it was the only acceptable answer.

"Do you know what day this is?"

"It's Saturday."

"What date is it?"

"Oh. Uh, I think it's the 17th."

"It's September 17th, and all you can say is that you think it's the 17th?"

Now I was getting confused and even more intimidated.

"Boy, this is Hank Williams's birthday!" he shouted.

"Oh, uh, well, happy birthday, Hank," I said, hoping to ease the tension.

"January 1st, 1953," he said in a voice that was much quieter and less intimidating.

"What?"

He opened another bottle of OB. "I was here in Korea, and I had just come back from 30 days on the front lines," he said, then took the first swallow from the new bottle. "I was going through the chow line, and somebody told me. 'Sarge, you hear about Hank Williams?' What about 'im, I asked, other than the fact that he's the best singer that ever lived? 'Then you ain't heard? Last week, on New Years Day it was, Hank Williams was goin' to a show, 'n' he died in the back seat of his car.'"

The master sergeant's voice had choked up as he delivered that last bit of information.

"'N' boy, what I done then, 'n' I ain't none ashamed of it neither, is I cried, right there, in the mashed taters."

I stared at him, looking for something to say, then muttered, "I'm sorry."

He lifted the bottle of beer. "That's all right. Ole' Hank, he's singin' in heaven now, 'n' them angels ain't never heard nothin' no purtier."

I nodded, then walked away to the opening strains of "I'm So Lonesome I Could Cry."

Corporal's Stripes

When I was in Korea, I was offered the opportunity to convert from the E-4 rank of specialist to corporal, if I would agree to become NCOIC (noncommissioned officer in charge) of the fuel dump. I had six Korean civilians working for me, but I had no soldiers.

Normally that wasn't a problem, but one cold winter night it became a huge problem. I was sound asleep in my bunk in the hooch when the CQ came to awaken me. I groggily opened my eyes, hearing the sleet pounding against the corrugated sheet metal that formed the Quonset-hut barracks.

"Corporal Vaughan, Colonel McKenzie just called the orderly room. He is out of fuel, and his heat has gone off. The OD wants you to take care of it."

I could not have heard worse news. It wasn't just that I would have to go out in a sleet storm in below-zero weather, I was going to have to refuel his tank. And because of the way the fuel tank was mounted, it was impossible for me to do that without help. I couldn't get my Korean civilians, as they all lived in the village. In fact, I didn't even know where they lived.

I looked at the nine other bunks, at the men nestled down in their blankets. Dare I start waking them up, one at a time, asking—no, begging—one of them to help me?

There was a new man in the company who had arrived less than a month earlier. His name was Bates, and he was a private first class. The only reason I knew him was because we were in the same barracks. We were friendly but not friends, in that we didn't hang out together. We were in different sections and rarely saw each other during the day.

It turned out that Bates happened to be awake and overheard the CQ telling me what needed to be done done.

"That's not something you can do by yourself, is it?" Bates asked.

"No."

Bates sat up in his bunk and began pulling on his socks. "I'll help."

I watched Bates as he began pulling on his OGs (winter fatigue uniform). I hadn't even gotten up the courage to ask anyone for help, but he had volunteered. I was so shocked by his offer that I just stood there; making no effort to get dressed.

He looked up at me and smiled. "You're goin' to get awfully cold if you go out like that."

It was 0300, and the motor pool, where we would have to go to get the POL (petrol, oil, and lubricants) truck, was a three-quarter-mile walk through the cold and sleet. When we got there, the truck wouldn't start, so we had to get another one and back it up to the POL truck, then roll a 435-pound barrel of fuel oil from one truck to the other. We also had to move the hand pump to our replacement vehicle so we'd be able to pump the fuel from the 55-gallon drum into the colonel's 20-gallon drum. It was attached to the bed of the truck, so we had to find a tool box to take it down, which took another 15 minutes.

By the time we reached the colonel's hooch, it was nearly 0400, and Bates and I were practically frostbitten. Bates climbed up onto the stand to stick the nozzle down into the tank, and I operated the pump. Colonel McKenzie stepped out of his hooch, bundled up in his parka.

"You are a couple of good men to come out on a night like this," he said. "I want you to know how much I appreciate it."

"Thank you, Colonel," I replied.

Colonel McKenzie's appreciation, sincere as it was, couldn't possibly equal the appreciation I felt for PFC Bates, who not only volunteered to help but did so cheerfully, and without the slightest complaint. All right, Bates didn't risk his life to save mine, he didn't run through enemy fire, he didn't do anything that rightfully earns brave men medals. But there is no doubt in my mind that, given such a challenge, a man like Bates would rise to the occasion.

A Christmas Carol

It was Christmas Eve of 1960, and I was in Korea. I was lying in my bunk with a broken foot, the result of a 55-gallon drum of oil having fallen on it a couple of days earlier. It was bitterly cold outside, but through the window of the hooch I could see a Santa Claus mask dangling from the concertina wire that surrounded the compound.

I was listening to "A Date with Diane" on the radio. I don't even know if Diane was her real name, but she was a disc jockey for AFKN, and every GI in Korea was in love with her. It wasn't just her sultry voice and the suggestion of intimacy, as if she were speaking to us individually. There was more to it than that. She reminded us of home: wives, girlfriends, the girl next door.

I was feeling particularly down this Christmas Eve. I was practically immobile because of my broken foot, I was 7,000 miles from home, I had a two-year-old son who was growing up without me and an eight-month-old son that I had not yet seen. And on the radio, Diane was playing for at least the tenth time, "I'll Be Home for Christmas." I knew that the only way I would be home would be exactly as the song says: *only in my dreams.*

The barracks was empty, all the other men having gone to the mess hall for Christmas dinner. I didn't feel like trying to make the walk through the cold, but a couple of men had promised to bring me a sandwich. I lay there thinking about Christmases past: getting up before dawn on Christmas morning . . . the smell of ham in the house . . . candy, nuts, oranges . . . the bicycle I got when I was 10 . . . the catcher's mitt, shoulder pads, and football helmet when I was 11 . . . a double-barrel, 20-gauge shotgun when I was 12.

Now I lay there listening to the wind whistling by the Quonset hut's corrugated tin covering . . . and, of course, the radio. I was getting a little hungry and regretted not trying to go to the mess hall. One thing about the army, they always fed us well on Thanksgiving and Christmas.

"We wish you a Merry Christmas, we wish you a Merry Christmas. . . ."

The song wasn't coming from the radio but from outside. The hooch door swung open, and nine men came in, singing and carrying trays filled with food.

"Vaughan, if you can't go to Christmas, we'll bring Christmas to you," Gus Thomas said. Footlockers were moved to the center of the hooch and converted into a table. One of the men had a small crèche that he had borrowed from the orderly room, and that became our centerpiece. The food was laid out, and we ate heartily, sharing stories of Christmases past and our plans for those yet to come.

As I look back on that day, I think I can honestly say that it ranks up there with one of the best Christmases ever. I enjoyed a camaraderie with friends who, at the time, were as close as family. No, they *were* family, and though I have never seen any of them again since that Christmas, they occupy a place in my heart that will forever keep their memory green.

I Make Sergeant

Shortly after returning to the States, I made sergeant. I was now NCOIC of the Prop and Rotor section of Fort Rucker's aircraft maintenance school.

I was in my office working on a lesson plan when one of my men stuck his head inside the doorway. "Sergeant, there's a guy out here who says he's your brother."

"What?"

I went out onto the hangar floor, and there was Tommy, in uniform but without his shirt. When my brother went into the army, he became a paratrooper, but his devil-may-care attitude wasn't compatible with army regulations. Eventually, after serving his time, he got an honorable discharge, but he never rose above the rank of PFC.

"Tommy, what are you doing here?"

"I was pulling an extra duty detail at the airport, and I heard someone say they were flying to Fort Rucker, so I bummed a ride."

"When are they going back?"

"They've already gone back, but I stayed so I could

visit you."

"Tommy, that's the dumbest thing I've ever heard of. How did you plan to get back?"

"You're a sergeant. I figured you could get me back before I'm AWOL."

"You are already AWOL. Get your ass in the car; I'm taking you back to the airfield."

"You can't talk to me like that. I'm your brother."

"You are also a private and I'm a sergeant. Let's go."

Luck sometimes shines on people who have no fear, and it shined on Tommy that day, because I found a couple of officers who were about to go to Fort Bragg in a twin-engine L-23, and they agreed to take him.

"I'm Colonel Hastings," the pilot said.

"I'm Tommy."

I cringed.

Later, when I became a warrant officer, my parents came to see me get my bars pinned on.

"Huh," Tommy said to them. "I made PFC four times, and you never came to any of my promotions."

After Tommy left the army, he graduated from college, then went to FBI school but quickly became a DEA agent. True to form, Tommy became an undercover agent, which was very dangerous—if he had been discovered, he could have been killed. I told him once that he had the best possible job because he could hang out with low-life scum buckets in the most disreputable places, and if he hadn't been a DEA agent, he would have hung out with low-life scum buckets in the most disreputable places anyway.

Tommy has been gone for 30 years now, but I believe that God holds all in His memory, and all I have to do is think of Tommy, and he is there, just on the other side of my memory. At this moment, he is racing down the garage roof in a wagon with an ironing board wing and a huge smile on his face.

He was my brother, and I miss him.

Foam, Sweet Foam

As instructors, we wore an orange band around the top of our billed caps, which were stiffened, rather like the cap General Charles de Gaulle wore. That orange band gave us a certain amount of *cachet*.

The instruction took part in two ways: Platform Instruction, which required lecturing, interrupted every 50 minutes by a 10-minute break, and Practical Exercise, which was hands-on maintenance, working on the actual aircraft. I much preferred the lecture, for two reasons. One, I was pretty good at it. And two, I was lousy at "hands on."

Here is an example. I taught how to use the magnaflux machine, which was a device that could determine internal cracks in metal. I understood the concept perfectly and was very comfortable teaching it. But I had never actually done it.

Hayes Aircraft was the civilian firm that maintained all the aircraft at Fort Rucker, and they got a new magnaflux machine. Nobody at Hayes knew how to use it, so they asked Major Knox if he would send someone over to run some tests for them.

"Sergeant Vaughan, I got a request from Hayes to send someone out to run some tests on the magnaflux machine, and I figured who better to ask than the man who wrote the lesson and teaches the class."

I was terrified at the thought, then I got an idea. "Major Knox, I'll give up my Saturday morning. Why don't you ask them to send some people in, and I'll teach them how to use it."

Major Knox thought that was a great idea, so he suggested it to Hayes, and they also thought it was a good idea. I taught 10 of their civilian employees how to use it, and they sent a letter of thanks to Major Knox, who wrote a letter of commendation for my 201 file.

We kept our training aircraft in a hangar, and once a month we would have a fire drill. On one such drill, Sergeant Allain was supposed to run to the big button that would, in the event of a real fire, activate the foam generator, which would flood the hangar with foam and either extinguish the fire or prevent it from spreading any further.

Sergeant Allain ran to the big button as the rest of us moved the students to safety and/or grabbed up fire extinguishers to simulate using them. But Allain got so carried away that he actually hit the button, and foam began spewing out onto the floor from outlets all over the hanger.

"I'm sorry! I'm sorry!" Allain yelled. "I don't know what made me do that!"

The officers in charge were understanding, though classes had to be suspended while we cleaned out the hangar. It took two days to get everything back to normal, then a report of survey was done to justify the cost of the inadvertent discharge of the foam, and paperwork was submitted to recharge the generator.

Once everything was back in place and classes resumed, inspectors from the IG office came out to the hangar to investigate what had happened. The section officers—Major Knox, Captain Pritzel, Lieutenant Carter, and Chief Warrant

Officer Stover—were all present, and they called Sergeant Allain over to the actuating button so he could explain exactly what he had done.

"Well, sir, we were having a fire drill, and my job—in case it was a real fire—would have been to hit this button. And I don't know why, but I just hit the button . . . like this."

And yes, Allain hit the button . . . and again foam began streaming out into the hangar.

"I'm sorry!" Allain shouted. "I'm sorry!"

You could barely hear him over the sound of the generators and the laughter of the men.

Army: Warrant Officer

Call Me Mister

I was a warrant officer! My first tour of duty in this new rank was the 101st Airborne Division at Fort Campbell. When I arrived, I reported for duty at Battalion Headquarters, and the clerk called the company to send someone after me.

A sergeant, who reminded me of Burt Lancaster as the first sergeant in *From Here to Eternity*, approached and saluted. Wow! I was being saluted!

"Mr. Vaughan, I'm Sergeant Haverkost. If you'd come with me, please."

"I've got my car. I'll follow you."

"The CO wants me to show you around. Leave the car; we'll come back for it."

"Yes, sir . . . uh sergeant," I said, correcting myself quickly.

Sergeant Haverkost smiled but said nothing. As I got into his car, he opened the glove compartment and pulled out a pint of whiskey.

"Here's to you in your new rank, sir." He took a swallow and handed the bottle to me.

What should I do? Should I report him for drinking on

duty? Was he testing me?

Nervously, I took a drink.

"This is the barracks, sir," he said, parking in the lot. I followed him into the barracks.

"Ten-hut!" somebody shouted, and I took a brace at attention.

"Uh, sir, that was for you," Sergeant Haverkost said with another smile. Then he took me to the company HQ, which was in one of the hangars where the aircraft were brought for repair.

"Sir," I said, reporting to the company commander. "Warrant Officer Vaughan reporting for duty."

"I'm glad you got here, Mr. Vaughan. I've been waiting to go on leave. You'll have the company while I'm gone."

"Sir? What do you mean I'll have the company?"

"I mean you will be the company commander in my absence."

"Sir, you can't mean that. I mean, I just got my appointment. Wouldn't your senior officer do this?"

"You are my senior officer," Captain Bailey said. "You are also my only officer. Don't worry about it. If you run into any difficulty, see Sergeant Haverkost or First Sergeant Dawes."

Two days later, I did encounter a problem, and I went to Haverkost, because I was intimidated by the first sergeant.

"Sergeant Haverkost, Private Lemon was disrespectful to me this morning, and the truth is, I don't know what to do about it. I mean, should I defer this to the CO?"

"I'll take care of it for you, sir."

About 15 minutes later, Havercost stepped into the CO's office that I was currently occupying.

"Mr. Vaughan, Lemon has somethin' to say to you."

"Yes, by all means, tell him to come in."

Lemon stepped up to the front of the desk, saluted, and reported to me. "Sir, Private Lemon reporting as ordered, sir."

For a moment, I was too stunned to even return the salute. Lemon had a black eye and a swollen lip.

"Tell him," Haverkost ordered the private as I finally returned his salute.

"I'm sorry I mouthed off to you," Lemon said.

"Sir," Haverkost reminded.

"Sir," Lemon added.

"And?" Haverkost asked.

"And, uh, I won't do it again, sir."

"Very good, Lemon, you are dismissed" I said.

"That take care of it for you, Mr. Vaughan?" Haverkost asked.

"Yes, Sergeant. Uh, thank you."

"Anytime you need me."

Haverkost left the office, and I lay my head on the desk. Captain Bailey couldn't come back soon enough.

A Cigarette Roll

The 101st, which was my first assignment after becoming an officer, was an airborne division at the time. Most of the soldiers of the 101st were airborne qualified, which meant they made parachute jumps out of airplanes. I was in the minority, in that I wasn't qualified—nor did I want to be.

They call such non-jumpers "legs," or more accurately, "laigs."

My supply sergeant, Sergeant Gibson, was not only airborne qualified, he had jumped on D-Day. He was a good supply sergeant . . . when he was sober. And that was the problem. Sergeant Gibson was a drunk. But how could I discipline someone who had three actual combat jumps, including the big one on D-Day?

One day we received a new five-kilowatt generator, and I asked Sergeant Gibson to log it in and get it put where it needed to be. He appeared to be working, but by now I had been around him long enough to know he wasn't working but was drunk and just doing useless forms.

Two days later, the generator had not been moved, so I

had the two clerks put it on Sergeant Gibson's desk. When Gibson came in the next morning, he said nothing about the generator but took care of his paperwork in the four or five inches of desk left to him.

This went on for a couple of days, and I just left the generator there. Five days after I had the generator placed on his desk, he stopped what he was doing, tapped on the generator, looked over toward me, and said, "Generator."

"Yes, Sergeant Gibson, it's a generator," I replied, happy that he had, at last, noticed it.

"On my desk."

"Yes, on your desk."

"There is a generator on my desk."

We were getting somewhere now. He had spoken a complete sentence.

Gibson went back to doing whatever it was he was doing, and the generator remained on his desk.

Two days later, I had the supply clerks move the generator, and I logged it into the property book.

For two more days, Sergeant Gibson continued to work on the five or six inches of the desk that had been his work space for a week. Then he looked around his desk, stared for a moment, then reached out into the thin air and moved his hand around.

"Generator's gone," he said.

"Yes, I logged it into the property book."

"Oh," Gibson said as he reclaimed the rest of his desk.

A few weeks later, there was a scheduled parachute jump that the men called a "pay jump." They had to jump once every three months to draw their jump pay. The next day, about five of the men were gathered in my office, discussing the jump.

"Beats anything I've ever seen," one of them said excitedly.

"Well, I guess you would call it textbook," another said. "Perfect, absolutely perfect."

"What are you all talking about?" I asked.

"A cigarette roll. That parachute wrapped so tight I didn't think he could get it undone. Remember, we were jumping from only twelve-hundred feet. You wouldn't think anybody could react that fast."

"I need a pay jump," Sergeant Gibson said.

The others looked at him in surprise.

"You just jumped yesterday, Sergeant."

"I did?"

"You don't remember?"

"Oh, yeah, I guess I do. You say somebody got a cigarette roll?"

The men looked at each other in disbelief.

"Yes, Sergeant, somebody got a cigarette roll."

"Who?"

"You, Sergeant. Your chute was wound up tight as a rope."

"Oh," Gibson said without further comment. He went back to doing whatever it was he did all day.

An Unpleasant Surprise

Our company was undergoing a Command Maintenance Management Inspection, which was so intense that even the married officers were sleeping in the hangar. After five days of pressure, we passed the CMMI, and everyone returned home exhausted.

As an officer, I was authorized post quarters, so I drove back home ready to just unwind. The house was empty, which didn't concern me. One of Grace's friends was about to have a baby, and Grace had promised to be there for her when the time came. Apparently, she also took our two boys with her.

I lay down on the bed to take a nap. I was beat, and it was well after dark when I finally woke up. I opened the closet for something, and all of Grace's clothes were gone. The boys' clothes were gone, too.

A few minutes later, the telephone rang. It was Tom, Grace's father.

"You are now divorced," he said.

The divorce papers arrived about three days later. It was final, and like our wedding seven years before, I had no say about it.

Winning in Las Vegas

I was playing draw poker in Las Vegas, the gambling capitol of the United States. I was a little behind, but considering how long I'd been playing, I wasn't too dissatisfied.

After the ante, I drew the five cards of the initial deal: an ace of hearts, a jack of hearts, and a ten of hearts. I don't remember what the other two cards were, but they didn't help my hand.

"Cards?" the dealer asked, holding the deck out.

"Two," I said, thinking it would be great to get two more hearts, or perhaps two more aces, jacks, or tens. But even an additional ace would be acceptable, as the previous hand had been taken by a pair of kings.

The cards were dealt, and I picked them up, looked at them, then tried to breathe. I had drawn a king and a queen—*of hearts*. I was sitting in a poker game in Las Vegas, and I was holding a *royal flush!*

Now, the challenge was in how I would play this. I obviously wanted to get the pot as high as I could, but I had to be careful with my raises, as too big a raise would drive the others out.

The first round generated more raises, and the second round, too. The pot was growing larger, and I wondered if I could get one more raise from someone. All but one player dropped out, and he eyed me confidently as he countered my raise with a raise of his own. We raised each other back and forth until finally I called.

Triumphantly he lay down three queens and two tens—a full house. With a huge smile, I stunned everyone as I revealed my royal flush.

The pot had grown considerably, but before I could rake it in, two Las Vegas policemen came over to the game.

"Are one of you Mr. Vaughan?" one of them asked.

"I'm Mr. Vaughan."

"Do PFC Gilbert and PFC Lemon work for you?"

"Yes."

"We have them in jail. If you will come sign for them, we'll let them go."

"What did they do?"

"You brought a convoy through town, didn't you?"

"Uh, yes, sir, I'm sorry about that. I misread the strip map." I shuddered as I recalled leading a convoy of 15 army vehicles right through the middle of town. We were unable to maintain convoy integrity because of the stoplights, and for several minutes after I arrived in the lead jeep, three-quarter-ton and deuce-and-a-half-ton vehicles came straggling in.

"Gilbert and Lemon were driving a fuel truck," the policeman said. "They pulled into the Motel 6, took off their shirts and boots, and jumped into the swimming pool. Like I say, if you'll come sign for them, we'll let them go."

"All right, as soon as I rake in my winnings."

I looked at the other players, who had been following the drama with intense interest. Then I scooped up the pot, which was by far the biggest of the whole game.

"Mr. Vaughan, if I was you, I'd bust both of 'em," Sergeant Martel said.

Oh, I forgot to mention . . . yes, I was in Las Vegas, but I wasn't playing cards in one of the casinos. We were playing in an aircraft hanger as the 101st Airborne Division was about to take part in training exercise Desert Strike in the Mojave Desert.

Oh, and the size of the pot I had just won? $3.25.

The Skeleton of Mojave

The 101st Airborne took part in a joint training exercise in the Mojave Desert called Operation Desert Strike. One of our helicopters went down on top of a mountain, and I took a crew up there to recover it.

There was an old abandoned mine, and the men decided to explore it. A few minutes later, one of them came back all excited and announced that they had discovered a skeleton! I went in with him, flashlight in hand. All the clothes had rotted away, leaving only a belt buckle and the man's eyeglasses, which were completely opaque.

We called the sheriff, who came out and examined the skeleton, then pronounced that it could have been there for more than a hundred years.

I've thought about that often since then, and though I've never written a story about it, I've considered doing so.

A Date with Linda

There are very few, if any, officers' clubs anymore. I'm sure that some "advisor" somewhere made the politically correct decision to end an institution that had stood the test of time for the military all over the world. There was a degree of pride and accomplishment to be able to go to the Officers' Mess, and indeed, when overseas, the officers' club (aka Officers' Open Mess) is where we took our meals.

It was a place to come to and relax after a hard day, a place to visit with friends, and a place to have a drink without having to worry about any "undesirable" elements, whether in the States or overseas. It was also a place for official social occasions, and to be honest, that may be the part of the officers' club activities that I'd be sure not to miss if I were still on active duty.

When I was at Fort Campbell, the 101st Aviation Battalion would have a "command social" event about once a month. The battalion commander would send out a circular to which you'd have to respond to such questions as: RANK (for the assessment, the higher the rank the more it cost) and LOBSTER OR STEAK: Choose one.

I was divorced and didn't want to go to a party where all the other officers and their wives would be socializing, leaving me the odd man out. So I wrote on the bottom of the circular: *Will Not Attend.*

The next morning, I was summoned to the battalion CO's office.

"Mr. Vaughan, 'Will not attend' is unacceptable. You *will* attend."

After duty hours that evening, I went to the bar across from Gate Six. An attractive young woman named Linda worked at the bar . . . and when I say "worked at," I don't mean that she was an employee.

I asked Linda if she'd like to attend a party at the officers' club, and she said, "Are you serious? Yes, I'd *love* to go, and I won't even charge you. I've never been to the officers' club."

When I picked her up the night of the party, she looked really nice . . . not in her barroom way, but really nice. She enjoyed herself immensely at the party, dancing with many of the other officers, including the battalion CO.

Early Monday morning, I was again summoned to the CO's office.

"Mr. Vaughan, I need to talk to you about the woman you brought to the party Saturday night."

"Yes, sir. Isn't she pretty? Bless her heart, I think this was the most exciting thing that ever happened to her."

"How well do you know her, Mr. Vaughan?"

"Why, I know her very well, sir. In fact, I'm thinking about asking her to marry me. You think maybe I could arrange a military wedding? You know, with arched sabers outside the chapel?"

"*Mr. Vaughan, no, my god, no!*"

"Sir?"

"You don't want to marry a girl like that. Don't you know about her?"

"I know she has a young child, but that doesn't bother me."

"Mr. Vaughan"—and here the colonel paused to pinch the bridge of his nose while he composed himself. "Mr. Vaughan, I hate to be the one to tell you this, but I must, to keep you from making a big mistake. I have it on very good authority that the woman you brought to the party, and I'm calling her a woman instead of a lady by design . . . that woman is . . . well, there is no way to tell you this but come right out and say it. She is a prostitute."

"What? Colonel, are you sure? Where did you hear such a thing?"

"I'd rather not say. But I consider the information to be accurate."

"A prostitute? I had no idea!" Bowing my head, I mimicked the colonel's earlier act and pinched the bridge of my nose. Then, in the most pained voice I could muster, I said, "I . . . I've already told my mother."

"Mr. Vaughan, surely, knowing this, you don't still intend to go through with this, do you?"

"No, sir, not after learning this about her. I can't tell my mother this over the phone, I'm going to have to drive to Sikeston to tell her in person. It's only about 150 miles. I'll drive over there and back tonight."

"I know this has come as quite a shock to you. I tell you what, there's no need for you to make that trip in one night. Take a couple of days off."

"Thank you, sir. I'm going to have to come up with some other reason. I can't tell my mother that Linda is a prostitute."

"I'm sorry, Mr. Vaughan. I'm truly sorry."

Leaving the CO's office, I headed to the bar and was having lunch when Linda came in.

"What are you doing here in the middle of the day?" she asked.

I grinned broadly. "The colonel told me you're a whore and gave me a day off to recover from it."

Linda laughed. "Lieutenant B. must have told him."

Custer's Own

Leaving the 101st and Fort Campbell, I went to Germany, where I was assigned to the 7th Cavalry. I had heard of the 7th, of course, and George Armstrong Custer. Who hasn't? In fact, Custer is second only to Lincoln, in American history, for the number of books and articles written about him.

When I arrived, Colonel Roxbury told me, "You are the junior officer in the squadron, Mr. Vaughan, and that means you will be the historical officer."

I had never heard of the position of historical officer. But, I was briefed by the outgoing historical officer, who was glad to be rid of the job. I learned that we had a lot of Custer memorabilia: a pair of his gloves, his saber, some old 7th Cavalry ensigns, and—most interesting to me—a field log book kept in Custer's own hand of one the trips to the field made by the 7th. He devoted nearly an entire page to describing how bad the mosquitoes were and the ineffective ways they had of combatting them. I was holding a book that had Custer's personal writings. His penmanship, by the way, was very good.

Sometime later, as historical officer, I had occasion to look up the oldest former member of the 7th who was still alive, and I found Colonel Frank Andrews, who had been with the 7th Cavalry during the Spanish American War. We invited him to Germany, all expenses paid, for Organization Day. He said he was too old to make the trip, but he sent me a six-page, single-spaced, typewritten letter about his memories. To my amazement, his father had served with Custer, and Colonel Andrews could remember sitting on Custer's lap during some of the musical productions they had at Fort Lincoln. For me, corresponding with someone who could actually remember Custer was like reaching out to touch history.

I was in D Troop of the 3rd Battalion of the 7th Cavalry Regiment of the 3rd Infantry Division. The other two battalions of the 7th were in Korea and in the States, though I forget where. This was during the "between" time, between Korea and Vietnam, and the army had regiments split up and spread all over the world. D Troop was the only aviation unit of the 3rd Battalion.

I enjoyed being the historical officer so much that I volunteered to remain in that role for as long as I was in the 7th. In all my years in the army, this was my favorite assignment. The 7th had the most *esprit de corps* of any unit I had ever been affiliated with—we even had our own bagpipe and drum corps—and I really got into its history. Plus, who wouldn't like a long tour in Europe? We were inside the DMZ, which meant we were very close to the line that separated East and West Germany, and every time we flew somewhere, we would have to file an ADIZ penetration when we came back. If you forgot—as many of us did during our first few weeks there—you would suddenly look outside the helicopter and see an F-4 on either side of you—or perhaps a Luftwaffe F-104. Seeing an Iron Cross on the wings and fuselage of a German aircraft right beside you sometimes gave the eerie feeling that you might be in a B-17 some 20 years earlier. We were in Schweinfurt, after all, the scene of the great Allied bombing raids on their ball-bearing factories.

Why I'm Not a Drunk

On a different note, my time in Germany is why I am now a teetotaler. The event that brought it on was a change of commanding officers of the 7th Cavalry. We were celebrating at the officers' club in Schweinfurt with a change-of-command party. Many of us—including me—were participating in the drinking game of Thumper. The party started on Friday night, the 10th of January. The next morning, we were to have a change-of-command ceremony that would feature the troops marching in review, with helicopters doing a low-level flyover. But at 0430 on Saturday morning we were still thumping and drinking, when someone said, "We have to fly in the morning. We're doing a low-level pass over the change of command parade."

"It *is* the morning," someone said. "Too late to worry about it now."

Deciding it was too late to go to bed, we stayed awake but cut off the drinking. At 0900 we went to the airfield and climbed into the six helicopters that would make the pass in two "V" flights of three.

Our flight leader called Schweinfurt Tower and mumbled, "Shwmnfefl flgn taklf."

"Aircraft calling Schweinfurt Tower, say again, please."

"Schmenffl fnn tkbf."

"I can't read you. Please say again."

"*Schmmpfn tar flh takv!*"

Then the one officer who had not stayed up drinking with us all night called the tower. "Schweinfurt Tower, flight of six on the pad for takeoff."

"Roger flight of six, winds calm, altimeter two niner, niner two, cleared for immediate departure from helipad."

"*Raklensgmph fmbn that's what I said!*" our leader managed to get out.

I was in a Huey, on the right wing of the trailing V. Mike Lindell was flying, and as we lined up to make the flyover, he said, "Hey, Dick, bet I can make a cyclic over the *Rathaus*."

"I'll bet you ten dollars you can't."

"Watch."

The problem with attempting a cyclic climb is when you pull the stick back you get sort of a climb, but it is counter-productive, because you flare and slow down. But Mike tried it.

"You aren't going to make it, Mike. . . . You aren't going to make it. . . . *Mike, you're not going to make it!*"

Mike moved the cyclic forward and pulled up on the collective just before we crashed into the city hall building.

"Chief, we hooked the weather vane up on the skid," the crew chief said laconically.

The right skid hooked onto a weather vane on top of the *Rathaus*, pulled the vane off, and dropped it into the Main River. The weather vane was 400 years old and had been carefully protected during the ball-bearing raids during WWII. We did, with one helicopter, what a thousand B-17s had not been able to do during the war: We destroyed that weathervane.

When I sobered up and realized what we had done, I was so chagrined that I swore, on that very day, I would never drink again. It is now 56 years that I have been faithful to that vow.

Macho Pilots Brought up Short by a Little Old Lady

One day when weather suspended all flight operations, a bunch of us were in the hangar, getting out of the rain. There were some schoolteachers there as well, single American women who had come to Germany to teach the dependent children of the American soldiers.

As a lot of pilots are wont to do, we were "hangar flying." And, with an audience of single females, we were lording it up a bit.

"You know, twenty years ago, Schweinfurt was a pretty hot place," one of the officers said, and for a few minutes we talked about the air raids on Schweinfurt to stop production of ball bearings.

"If I had been around then, I would have been a B-17 pilot," one of the men said.

"Yeah, that B-17 was quite a plane. It would take a lot of hits and still fly."

For the next few minutes, we talked about the virtues of the B-17 and how we would been perfect for the missions.

All of the schoolteachers were listening to us in rapt

attention. One of the older teachers spoke up then.

"The B-17 wasn't that easy to fly," she said. "You had to fight the cowl flaps to keep the engines cool, and you had keep changing the trim to keep the props in synch. And I swear, in rough weather, it took two men and a small boy to keep it steady."

"Yeah? What do you know about it?"

"I was a ferry pilot in the war. I have a little over 800 hours in B-17s."

We shut up then, because there was nothing else we could say.

Earning the Bird

GCA—ground controlled approach—is a system whereby a radar operator on the ground can talk a pilot down through near zero visibility. During the Berlin Airlift it was used extensively and no doubt saved a lot of lives. It was also used throughout military, commercial, and general aviation. With today's ILS and such systems, GCA is no longer used, but when I was in Germany it was still in effect. Though I never had to actually use GCA, we'd sometimes get a request from the ground asking if you would accept a practice GCA.

One day I was approaching Fulda and called the tower: "Fulda Tower, Army 715, downwind for 27."

"Army 715, call base."

"715, turning base."

"715, will you accept a practice GCA?"

"Affirmative."

I waited a second, then a female voice came on. "Army 715, GCA, squawk your parrot."

I activated the transponder, which sent out a signal so they could identify me.

"I have you, 715."

"715," I replied.

"Do not acknowledge any further transmissions, 715. Please establish a 500-foot-per-minute rate of descent."

I did so.

"Come left, zero five degrees."

I did.

"You are on course, on glide path, your approach is good. Maintain vector 27 and 500 feet per minute rate of descent. . . . On course, on glide path, your approach is good."

This continued for another couple of transmissions. Then I stopped, came to a hover, and started backing up.

"You are on course on glide . . . uh . . . on course . . . uh . . ."

She stopped transmitting, then a moment later I heard her panicked voice. "Sergeant Johnson, Sergeant Johnson!"

Next was a male voice. "Army 715, are you helicopter?"

"Affirmative."

"Discontinue GCA, land at pilot's discretion."

"715."

After completing the approach, I hovered by the tower to the helipad. I saw an angry young WAC (as we called them then) standing in the window. She "saluted" me . . . but it was NOT the kind of salute you would exchange in polite company, bless her heart.

Walk Like a Damn Duck

Those who know me, know that I "walk funny." By that, I mean that I tend to walk with my toes pointed outward. Slue-footed is the common term for it, though the correct term is supination. As a kid I was teased about this, and often the teasing was hurtful.

Once, while in the army, I had to attend a meeting in the operations shack at the airfield. I landed at the helipad just in front of the operations shack and walked across the tarmac. I could see, in the big window, several of the officers watching me approach. When I went inside, they were laughing.

"Why is everyone laughing?"

"Vaughan, has anyone ever told you that you walk like a damn duck?" Dan asked.

"Yeah, Dan, they have. But, when I was 12 years old I was in a terrible car wreck. Both my sisters were killed, and I was taken to the hospital, not expected to live. Finally, I pulled through the initial crisis, but the doctor told me I would never walk again. I wouldn't accept that, Dan. So I would fall out of bed trying to walk, and the nurses would have to put me back. Then I got to where, by holding onto the bed and the chair, I could go to the door and back. Then I tried the hallway, sometimes getting to the very end of the hall before collapsing from pain and exhaustion, and they would have to put me in a wheelchair to take me back to my room. But I didn't give up, Dan. I kept trying, and one month later, when my folks brought me home, I walked from the car to the house. I had to use crutches, but I walked. Then, I remember the day I walked all the way around the block without crutches. My family celebrated, but it was a bittersweet celebration After all, my two sisters weren't there to celebrate with us. By the time I was back in school, I was walking again, Dan. From the bus to the school, and from classroom to classroom without help . . . I was walking! Now, Dan, I know I walk funny, but I am just thankful that I can walk at all."

By now, in front of nine other officers, tears were streaming down Dan's face, and he wasn't the only one. I even managed to tear up myself.

"No, man, look," Dan said. "You walk good. You walk real good. I mean, you walk better'n me. I'm sorry, I didn't know all that."

I never told Dan or any of the other officers that there was no accident and I never had any sisters. I just naturally walk like a damn duck.

Slap-Face Dance

Oktoberfest was always fun, but I did something very dumb at one of them. I was in a tent with a group of my friends, drinking beer (in my "before I quit drinking" days) and listening to the German band play "oompah" music. There was a German sitting across the table from me, and to keep time with the music, we began clapping hands . . . first our own, then reaching across the table to clap each other's hand. We were doing this with such enthusiasm that we climbed up onto the table to continue this rhythmic clapping, with the impact of the clapping growing stronger and stronger.

Somehow this led beyond merely hitting each other's hand . . . to slapping each other in the face, still in rhythm with the music. We would clap our own hands, then slap each other's hand, then slap each other in the face.

"Dick, are you crazy? What are you doing?" my friend Mike called to me.

"I don't know, but I'm not going to stop until he does," was my really stupid answer.

The music continued, but by now everyone else had

stopped what they were doing to gather around our table and watch the German and me exchange blows. The band got into it, and they also moved down to the table. I was determined not to stop until either the music or the German stopped.

Neither seemed willing to do so, and we began hitting each other harder, still in rhythm with the music, perhaps thinking that if we could knock our adversary completely off the table it would stop. But it's not easy to knock someone off the table just using an open palm. And this had not developed into a fight. On the contrary, the German and I continued to smile at each other, somehow thinking that losing the smile would mean losing composure and perhaps even our manliness.

We continued to slap each other, the crowd clapping in rhythm and urging us on. I saw that one of the eyes of the German had swollen completely shut, but I could only see that with my right eye, as my left was also swollen shut. A couple of times one of our knees would buckle, but we maintained our footing and kept at it. After all, we were having so much fun.

Then, perhaps having mercy on us, the band stopped the music and we were able to stop. The German and I shook hands, wincing with pain because by now our hands were also swollen.

"Wow, I can't believe you did that," Mike said when I climbed back down.

"Why didn't you reach up there and pull me off the table?" I asked. But because my lips and cheeks were swollen, my question came out sort of like: "Y nt wret thr grob lel?"

Somehow Mike understood. "Because, man, that was awesome!"

When I examined my face the next day, I never realized you could get so many colors out of the human skin.

Obviously I occasionally think about that dumb episode—or I wouldn't remember enough of it to write it here. I wonder if that German remembers it.

A Coincidental Visit

I enjoyed taking leave to explore, not only Germany, but much of Europe. I loved the food and the "fests," and I very much enjoyed the camaraderie of the men with whom I was serving. I was in the Seventh Cavalry, and there was a mystique and an *esprit de corps* in the unit that was greater than any other unit I ever experienced.

Though I was a single officer while in Germany, dependents were authorized, so there were many American wives and children there. The children picked up the German language very quickly, and I used to enjoy being in a store in downtown Schweinfurt, watching an American wife deal with the store clerk through her child.

"Ask him if he has this in red."

"Mom, I'm bored. I want to go home."

"Ask him if he has this blouse in red."

"*Haben Sie diese Bluse in rot?*"

"*Ich habe es nicht in rot, aber ich habe es in gelb. Ich glaube, sie würde in gelb gut aussehen.*"

"He says he doesn't have it in red, but he thinks you'll

look good in yellow. Can we go home now?"

Once, when I was buying a sweater, I complimented the clerk on his excellent command of English.

"Oh, I love America," he said. "I've spent some time there. Where are you from?"

"Oh, a little town that I'm sure you've never heard of," I replied. "I was born and grew up in Sikeston, Missouri."

"Oh, I love Sikeston. I was stationed there during the war."

If he had hit me in the stomach with a baseball bat, I wouldn't have been more stunned.

"What?" I asked, incredulously. "Did you just say that you were stationed in Sikeston during the war?"

"Yes."

"What do you mean you were stationed there? We had an Army Air Corps training base there, but I thought you were German. Are you telling me you were in the American Air Corps?"

"Yes, there *was* an air base there, wasn't there? I do remember seeing all the training planes flying around. I am German, but I was with some Italian troops when I was captured in North Africa. We were taken to America, and for the rest of the war I was a POW in Sikeston."

I would have never thought that being a prisoner of war somewhere would be the same as having been stationed there, but I suppose you could make that case. I do remember that there was a POW camp just north of Sikeston in a place we called Grant City, which housed Italian prisoners. I would often see them in town, easily identified because they wore khaki uniforms but had a big green oval on the back of their shirt, with the letters POW. They had an amazing amount of freedom of movement, and one morning each week, the swimming pool would be closed to the public so the prisoners could use it.

Although most of the prisoners were Italian, there were some exceptions, as in the case of Herr Papf, the clerk who sold me the sweater. If a German soldier was embedded with the Italians when captured, he could wind up in prison with them. One of the Italian prisoners, who happened to be from a very wealthy family, fell in love with and married a local Sikeston girl. After the war, she moved with him to Italy.

I saw Herr Papf several more times after I bought the sweater, and he always had warm regards for Sikeston.

"The best thing that could have happened to any German soldier during the war was to be captured by the Americans. Being in the American prison camp was like going to a college somewhere. We ate well, we had a lot of freedom, we worked on the farm." He laughed. "I picked cotton. When I tell my friends I picked cotton, they don't even know what that means."

"I picked a lot of cotton, too," I said. "Three-and-a-half cents per pound."

"We were paid, as well," Herr Papf said. "It went to the prison, not to us as individuals, but the prison officials used the money to improve our lot. I would like to go back some day."

I lost contact with Herr Papf a long time ago. I don't know if he ever made it back to Sikeston, but his memories were so strong and positive that I am certain he was able to revisit anytime he wanted, just by thinking about it.

I was prepared to extend my stay in Germany, but that was not to be. My application for extension was denied and I received orders for Fort Riley—to help form a company that was going to Vietnam.

At that time, I had been dating an American schoolteacher named Gail for three or four months, and when I got orders to leave Germany, we got married . . . I think. At least we went down to the *Rathaus*, where someone spoke a lot of words that neither one us understood, then gave us a paper that said we were married.

Fort Riley was the home of the old 7th Cavalry, and though I was only there for two months before we left, I soaked up as much of the history as I could. To any old 7th Cavalryman who might read this: "Garry Owen."

In Transit

During the Vietnam War, we deployed as individuals, which meant that despite there being as many as 200 on board the plane, you didn't know anyone. You were all alone.

If you were lucky, you got a window seat, and during the six times I crossed the Pacific—three over and three back—I was lucky three times. The first time was the most memorable. A few days earlier I had said goodbye to everyone I loved, and now I was at 30,000 feet looking out at the static electricity dischargers from the trailing edge of the wing, marveling at how the laminar flow kept them from fluttering in the 600-mile-per-hour wind.

The stewardesses—not flight attendants, stewardesses—were all pretty, exceptionally friendly, and sometimes a bit flirtatious, doing their bit to make everyone feel good. A few of the men were joking and laughing, perhaps a bit too hard . . . but most of the men were either lost in their own thoughts or talking quietly with the person next to them.

The food was always outstanding—filet mignon for lunch, roast beef for supper, great omelets for breakfast. We landed at Wake Island, and I couldn't help but think of the battle that took place here during WWII, when we lost Wake to the Japanese. As we circled for a landing, I was surprised at how small the island is. As an aside, I was on Guam three times before I realized they had daylight there, just like in the rest of the world.

I read *The Godfather* and watched the movie *Oklahoma* during that first flight over.

We landed at Tan Son Nhut in the middle of the night—0100—and were assigned a bunk in the receiving tent.

"Damn, looks like it's goin' to rain," one of the recent arrivals said. "Hope the tent doesn't leak."

"No rain," the processing sergeant said.

"What is that, then?" the man asked about the constant lightning flashes and the distant rumble of thunder.

"Artillery."

"What?"

"Don't worry, it's ours. Here's your mosquito net."

I had never put mosquito netting onto a bunk . . . and besides, it was the middle of the night and too dark to see. I lay the netting on the floor beside the bunk and went to bed—but not to sleep. I was unprepared for the ferocity of the mosquitoes. You didn't feel them when they bit you, so you couldn't slap at them. But within seconds, you were on fire from 15 to 20 to 50 mosquito bites.

It took three days to be in-processed, and it was a miserable three days, even with the netting. I didn't bother to make friends, as we were going to be spread out all over country. My mail had not caught up with me, and as yet I had no APO address and couldn't write home.

Finally the day came that a helicopter arrived from Phu Loi to pick me up and take me to my assignment. It was being flown by a smiling Bob Bivens, a warrant officer I knew.

I was home.

The 500 kW Generator

I was in Vietnam, and I had been given the collateral duty as PBO, or Property Book Officer. The frightening thing about being PBO is that you have to sign for every piece of TO&E equipment the unit has, so that you are responsible for several million dollars' worth of materiel. The good thing about being PBO is that you are often in position to make things happen.

When I arrived at Phu Loi for my first tour in Vietnam, our entire company area was without electricity except for a few small generators here and there. As the PBO and supply officer, I put in for a larger generator, only to receive the dreaded response: Back Order.

"We've got to have power here, Mr. Vaughan. Do what you can do to get us a large-enough generator to do the job," Major Royal said.

The next day I drove to Saigon and checked every supply depot there. In the USARV depot I saw two beautiful 500 kW generators sitting on a large trailer. I had a larcenous idea and returned to Phu Loi, where I forged a DA Form

1687 (receipt for equipment). I entered the unit ID as HQs 2nd Brigade in Di An, thinking that would lead any investigation away from the 605th at Phu Loi. I listed the receiving officer as Captain John Hazard, taking particular delight in choosing the name, as what I was doing was very hazardous. First, I was forging a document. Second, I was a chief warrant officer, about to pass myself off as an army captain. And third, I was going to steal a 500 kW generator, worth about $15,000 in 1966.

My next step was to take a shirt down to the Vietnamese tailor and have black bars sewed onto the collar, with the name Hazard above the pocket. Afterward, I explained to my supply sergeant what I was doing and gave Sergeant Loomis the chance to participate, or pass.

"Chief, can we steal a movie projector, too? I know the guys would really like to watch a movie at night."

"Do you know where one is?"

"MACV has ten of 'em just sitting there. Nobody's using them."

"Sure, why not?"

"Then I'm in."

"The problem is, I want to go up on Sunday when there won't be many people around. And with just two of us, I'm not sure how we'll get one of them from the flatbed into the back of a deuce and a half."

"Why don't we just take up a five-ton tractor and haul off the whole trailer, with both generators?"

"Sergeant Loomis, you're a genius!"

Sunday morning, with my temporary promotion to captain, we took the five-ton to Saigon. The first stop was our dry run at the MACV supply depot. If we had no trouble getting a projector, then it should work for the generators.

"You'll have to sign here, Cap'n," the gate guard said.

"I'll be glad to," I said, signing the name John Hazard, Captain, INF.

The gate guard picked up his copy of *Playboy* and waved

us through. Ten minutes later the projector was on the floorboard between my feet, and we were headed for the USARV depot.

"We're closed," the gate guard, a SP/4, said when we drove up.

"Specialist we drove down from Di An today, and we took fire. There's nothing on my 1687 that says I can only take delivery on weekdays. I don't know how it is down here in Saigon, but up country we fight the war seven days a week. Now, do you want me to go back to General DePuy and tell him he can't have his generators because it's Sunday? Do you have a phone? I'd be glad to call the general so you can talk to him."

"Uh, no, sir. You can go on in."

We drove into the huge depot and spotted the two 500 kW generators sitting on the trailer in the back corner. And as luck would have it, there was easy access to back the tractor under it. Working quickly, we backed under the trailer, raised the support gear, then drove off.

"Hold it!" the guard shouted as we passed through the gate.

"Oh, lord, Loomis, we've been found out. I'm going to Leavenworth," I said as my heart leaped to my throat.

The gate guard came up to the window and pointed to the trailer.

"Those are my tie-down straps!" he said.

Quickly, Sergeant Loomis jumped down,, climbed up onto the trailer, and removed the straps.

As an addendum to this story, the last time I saw Phu Loi was in 1972 on a subsequent tour. Both 500 kW generators were still there, now on concrete pads and occupying a place of honor inside their own sheds.

A Night on the Town

Once, I was pay officer in Vietnam, and I had to go from Phu Loi to Vung Tau to draw about $20,000 in MPCs (military payment certificates; we didn't use US currency) and about $5,000 in piasters (South Vietnam's monetary unit at the time). This was a double blessing. I would get to spend the night in Vung Tau, which was a lot more pleasant than Phu Loi. And stationed there was Mike Lindell, a warrant officer and very close friend I had served with in Germany, the same person who attempted a cyclic climb over the *Rathaus*.

I drew the pay and went to the orderly room to have them hold it in the company safe for me. But army regs said that the company safe wasn't secure enough, so I put all the money in Mike's laundry bag, and we went to town. We were having a lot of fun, drinking with the guys and buying Saigon Tea for the girls. Yes, I know this was Vung Tau, but Saigon Tea was the ubiquitous drink for the girls who worked at such places as the Pretty Girl and Happy GI Bar.

We ran out of money; after all, payday wasn't until the next day. But I was pay officer, so I returned to the hooch and got $20 worth of piasters, and the fun continued. When we ran out of money again, I made the piaster run a second time. By now, Mike figured out where the money was coming from, and he offered to make the trip. Two more times, and we were $100 into my piaster payroll. But we were having great fun—so much so that some of the bar girls were following us from bar to bar.

An old CW4 (chief warrant officer 4) who had served in WWII and Korea got curious. "Where's Dick getting all that money?"

"Oh, no sweat, man. Dick is the pay officer. He's got the whole payroll in my laundry bag back at the hooch," Mike said, innocently enough.

"What? *My god! We're all going to be court-martialed!*"

In pre-flight the next morning, I "found" a reason to keep us grounded long enough for everyone who had been in our bar-cruising adventure the night before to draw their pay and convert enough of it into piasters to make up the shortfall. I think the finally tally came to something just over $200, which was a considerable amount then.

Even though I replaced the money the next morning, I guess that technically I was guilty of embezzlement. But that was 50 years ago. Surely the statute of limitations has expired, hasn't it?

Gertrude

I'm a pretty good cook, if I say so myself. I enjoy cooking and do almost half the cooking in our house. But what I am particularly good at is barbecue. After my dad left the trucking business, he owned a market in Sikeston, and it got the reputation of putting out very good barbecue. He did Memphis style, which to me is the *only* barbecue.

When Ruth and I still lived in Sikeston, our church, St. Paul's, did a big barbecue every Fourth of July. I miss those barbecues. In addition to pork, I would also barbecue goat. Several people who swore they'd never taste goat decided that they liked it after all.

My first experience barbecuing goat was in Vietnam. Almost everywhere you went in the small villages outside of Saigon, you would encounter a *papasan* with six to eight goats. I decided to do one for the company, so one day I went to *papasan* to buy a goat. We began bargaining, but we weren't making much progress. I was angry because I was sure he was charging way too much; he was angry because I was offering way too little.

Then we had a communication breakthrough. He thought I wanted to buy all six of his goats, and when he realized I wanted only one, we were able to negotiate a price that satisfied both of us.

I put the goat in the back of the jeep and took it back to the company area. It was two weeks before the Fourth, so I tied the goat outside the maintenance hangar. I named the goat Gertrude, and the men all came by to feed and play with her. Then one day she got loose, but she didn't leave the company area. She wandered around, day and night, often spending the night in one of the hooches.

A few days later, Colonel May called me into his office.

"Mr. Vaughan, what do you plan to do with Gertrude?"

"I know she's getting in everyone's hair, Colonel, but it won't be long. I'll have one of the Vietnamese kill and dress her, then I'll barbecue her."

"You will not, sir!"

A colonel had called me, a warrant officer, sir. I knew he didn't mean it.

"Sir?" I replied. I did mean it.

"You *will not* murder Gertrude. How can you even think of killing someone whose name is on the company roster?"

"Sir, what are you talking about? You knew I planned to barbecue a goat for the Fourth. You gave me permission."

"Yes, but I didn't know you were going to *name the damn thing*! Now Gertrude is one of us, and as long as I command this outfit, she will not be harmed. Do you read me, Mister?"

"Yes, sir, loud and clear."

"Now, go buy another goat. But by the time it reaches the 56th, it had better look like something I could get at the meat market."

"Yes, sir," I said. The truth was, I had already been having second thoughts about killing Gertrude.

While I was cooking the new goat—which in accordance with Colonel May's orders could have come from Krogers or

Piggly Wiggly—Gertrude, who was a most curious animal, came over to see what I was doing.

"Gertrude, I hope this isn't anyone you know," I said. "Or even worse, someone from your family."

Evidently Gertrude was fine with it, because she held no hard feelings toward me.

Generally we returned to the States 12 months after we arrived, and since none of us arrived together, there was a constant rotation of all the officers. By the time my drop date rolled around, Colonel May had already gone home, and there were only three officers who knew the story of Gertrude and why she had freedom to go anywhere she wanted.

We had built a small officers' club between two of the hooches, and I was given a going-away party the night before I left. Gertrude was present . . . inside the club. It was only right. After all, I had been her sponsor.

Miss Sahn Offers to Buy Me an Airplane

One of the big drawbacks of being a PBO is that the property book officer is financially liable for everything that's on the book. Most of the time there's no problem with the big items—jeeps, trucks, aircraft, weapons—because they aren't likely to get away from you. It's the toolboxes, special tools, and smaller generators that disappear. Toolboxes, especially, have a way of getting lost. At the time, a toolbox with all the allocated tools cost $73. Some special tools could cost as much as $100, and the smaller generators were $300.

Periodically, the property book has to be reconciled so that what it says on the property book pages is what you have. But if they aren't on the property book page, then you aren't liable for them. It would seem that the solution would be to just create a new page, reflecting the number that you can actually account for. But there's a problem with that. A new page, so crisp and white in a property book that has pages that are two or three years old, stands out like a neon sign.

However, for the clever supply officer there was a solution. Make up a new page, reflecting what you actually have on hand, then use a shaving brush to lightly paint the new pages with coffee. It worked all the time.

One morning, after being unable to reconcile some of my pages, I had my secretary, Miss Sanh, type out new pages, then I told her to paint them with coffee. She had about seven pages spread on the floor and was painting them when Colonel May came into the supply room.

"Miss Sahn, whatever are you doing?"

"Oh, Mister Vaughan, he very, very smart, he can lie, cheat, and steal more better than anyone. If you paint the pages with coffee, they look very old."

Colonel May threw his hands up and, shaking his head, turned around. "I don't want to hear it," he said. "I don't want to know anything about it."

Miss Sanh, as it turned out, was much more than a secretary. I wasn't there very long before I learned that she owned at least a dozen buildings, including two that were being used as officers' clubs. We began making bets as to how wealthy she was, and the general consensus was that she was worth at least a million dollars.

One day, I was looking at a *Flying Magazine*, drooling over a Cessna 210.

"Mr. Vaughan, you like that airplane?"

"Oh, yeah," I said. "It's something you daydream about."

"If you divorce wife American and marry me so I can go to America, I'll buy that plane for you. Then we divorce and you marry American wife again."

As things subsequently turned out, I wish I had taken her up on her offer.

William and Mary

My greatest personal disappointment is that I don't have a college degree. Admittedly, at my age, there is no longer any need for a degree as far as work opportunities are concerned. But it is a matter of personal pride, a feeling of failure on my part, and if I can ever catch up with the books under contract and waiting to be written, I may explore the possibility of pursuing a degree.

Following my first tour in Vietnam, I was assigned to Fort Eustis in Virginia. While there, I had the opportunity to attend William and Mary. I knew I wouldn't be able to go long enough to get a degree, so I didn't bother to construct a curriculum that would lead to one. Instead, I took every undergraduate and even some graduate classes in English. I made straight A's in every class but one, English Lit, where I got a C.

The professor was a woman, and when we first started, I thought the class would not only be easy but enjoyable. We were to read *Catch-22* by Joseph Heller and *The Naked and the Dead* by Norman Mailer. I had met both Heller and Mailer and had read their books more than once.

Our assignment was to read and analyze the books. I wrote what I was certain was a good essay on *The Naked and the Dead*. And why wouldn't it be good? Mailer had based it on his personal experiences in combat, and I had been in combat and could relate. And, as I said, I had met Mailer, and we had discussed both our war experiences and writing. I confidently submitted my report and, at the next class, received the following from the professor.

"I am giving you a C because you have not been able to see through the superficial aspects of this book. You have perceived it as an adventure story; it is much more than that."

All right, I'll do better with *Catch-22*, a novel about flying and the antics of soldiers who must be just a little more insane than the war in order to survive the war. I had written *Brandywine's War*, which *The New York Times* called a "Very funny burlesque, comparable to *Catch-22*." Surely I could capture this one.

I got another C, with this response. "Again, you have failed to grasp the real meaning of this book."

Then I got a break. We were told that, for extra credit, we could read a book of our own choosing. I chose *Lust Empire*, which I had written under the pseudonym Dave Vance. Confidently, perhaps even a bit arrogantly, I submitted a report on this book.

Another C.

"You don't understand, do you, Mr. Vaughan? Perhaps you thought to shock me by choosing a book with such a risqué theme. However, even though this book would hardly qualify as literature, it is obvious that the author had an underlying theme when he wrote it. It is equally obvious that you have no idea what Dave Vance was trying to say. I am very disappointed in you."

I was too ashamed to tell her that I was Dave Vance.

Straight Arrow

We had a colonel who could have been the prototype for Herman Wouk's Captain Queeg in *The Caine Mutiny*. He was, to put it mildly, a very difficult officer to work for. Among his idiosyncrasies was his insistence to put his personal stamp on the battalion, and he chose to do that by giving the battalion his personal motto. As his model, he chose the 7th Cavalry, where officers and enlisted men greet each other by exchanging salutes and saying "Garry Owen." Why Garry Owen? It comes from an Irish marching tune, also spelled *Garryowen* or *Garyowen,* used for some military formations, including by George Armstrong Custer for the 7th Cavalry.

The 7th Cav may be the unit with the most *esprit de corps* of any regiment in the army. *Why,* the colonel must have asked himself, *shouldn't the 365th have the same spirit?*

The colonel decreed that every helicopter, vehicle, generator, air compressor, and even all correspondence bear his personal motto: Straight Arrow. To accomplish this, he had stencils cut, with a yellow arrowhead pointing upward, just above the motto. But instead of "Straight Arrow" in English, he asked a battalion clerk to find out how to express it in Vietnamese.

And therein was the problem. The clerk was less than a week from his drop date, so he acquired the information he needed from a bar girl with whom he had established a relationship during his time in-country. Then, he gave the phrase to the colonel . . . and from that day on, every piece of equipment and correspondence bore the motto *Mong Cua Ban*.

Soon after, we began to encounter strange reactions from the Vietnamese, who read the motto and laughed. Then the enlisted men began saying *"Mong Cua Ban"* as they saluted the officers. Sometimes they would say, *"Mong Cua* YOUR *Ban,* sir!"

The colonel was pleased that his idea had caught on so well, and he bragged about lifting the morale with the simple phrase. But it never dawned on him that the clerk was already home . . . and out of the army. And that the man had gathered the information *he* wanted.

You see, it hadn't taken long for us to figure out that *"Mong Cua Ban"* was Vietnamese for "Up Your Ass." The colonel didn't learn that until he returned to the States and read about it my book, *Brandywine's War*. Unfortunately, when he read my book, he was once again my commanding officer.

It did not go well for me.

My Prisoner of War

The recovery wasn't a difficult one; the lift-link wasn't broken, and the downed bird was intact enough that we had no problem rigging it to be sling-loaded out by Hill Climber or Box Car, the two CH-47 companies who were assigned to support us.

As the men were rigging the helicopter for recovery, I was passing time with the lieutenant who was in command of the infantry platoon that was providing us with security. There had been a battle there an hour earlier, but the VC had fled and the rest of the lieutenant's company was in pursuit.

Then, suddenly, a Vietnamese man in black pajamas and a conical hat stood up from the elephant grass, not 10 feet from us.

"Lieutenant, look out!"

The VC had an AK-47 in his hand, but he wasn't holding it in a threatening manner. He tossed it to one side, put his hands up, and shouted, *"Chieu hoi!"* He was giving up.

Quickly, some of the security platoon took him prisoner.

"Chief, I need you to take him back for me," the lieutenant said.

"Lieutenant, I'm not equipped to deal with a prisoner."

"I'm not either. I have to rejoin the company as soon as you leave. The last thing I need is a prisoner. I'm begging you, please take him back."

I agreed, and after the Chinook carried off the downed bird, my men climbed back in for the flight back. I asked my door gunner to keep an eye on our POW.

"Chief, if I'm going to watch this guy, shouldn't I have a handgun?" the door gunner asked a few minutes after takeoff.

That sounded like a reasonable request, so I looked over at my co-pilot, a brand new W-1, in-country less than a week. "Give McKay your pistol."

Walt nodded, pulled his pistol from his shoulder holster, then turned. McKay was too far away, so he handed the pistol to the VC prisoner, then pointed to McKay and mouthed the words, "Give the pistol to him."

"Holy crap!" McKay shouted.

The VC, with pistol in his hand, smiled, turned it around to hold it by the barrel, and handed it to McKay. He'd had enough of the war. . . . And to be honest, by that time, so had I.

But For the Grace of God

I was halfway through my second tour in-country, and it began as a routine recovery. A tail rotor pitch change link had been shot away, rendering the aircraft unstable. All things considered, the pilot did a good job of putting it down, and the injuries sustained by the crew were not life-threatening.

By the time we got there, the crew had already been evacuated by Dustoff (an acronym for Dedicated Unhesitatingly Service to our Fighting Forces—the call sign of the U.S. Army Air Ambulance units). All that remained was a pretty bent up UH-1C (the armed Huey, called hogs by the crews) and an infantry platoon to provide security for us, the recovery team.

An examination showed that the lift-link was undamaged, so it would be a fairly quick and easy job of rigging it for extraction. We didn't do the extractions; our job was to rig the helicopter so a CH-47 Chinook could come in and pick it up. And as expected, that part went without a problem. It was time to call in one of the big birds.

"Any Hillclimber, any Boxcar, this is Goodnature Three,

over?" Hillclimber and Boxcar were the call signs for the two Chinook companies in our operating area. We had a very high priority, so we always got a rapid reply.

"Goodnature Three, Boxcar Five One."

"Boxcar One, we have a downed Huey, lift-link intact, rigged and ready to go. Can you respond?"

"Authenticate, Goodnature."

"Authenticator is Vexation."

"Maddog," came the countersign. "Squawk your parrot."

"We activated the IFF (Identification Friend or Foe), which allowed him to home in on our signal."

"Roger the squawk. ETA zero-five."

"Hey, Chief, once he gets here, how long will this take?" the lieutenant in charge of the security platoon asked.

"You got a hot date L.T.?"

"Wish I did. I just want to get my guys outta here."

"He'll be hooked up and gone in less than a minute from the time he arrives."

"Chief, I see him," Sergeant Creech said, pointing south to where the tandem rotor helicopter was beating its way toward us.

"Boxcar Five One, we have eyes on."

"Pop smoke," Boxcar said.

Pop smoke meant that we were to toss out a smoke grenade so they could see where we were. The procedure was that you popped the smoke without identifying the color, then let the pilot tell what color he saw. That was because sometimes the VC or the NVN were listening in, and if you said you were going to throw green smoke, they would throw green smoke to try and lure the vulnerable Chinook in.

I picked up a smoke grenade and pulled the pin. *The grenade went off, even though I still had the spoon depressed!*

Yelling in pain, I dropped the grenade and put my hands to my face. Burning particulates filled both eyes, and I was instantly blinded. I took a couple of steps, then fell, only barely cognizant of someone yelling.

"Canteen! Get me a canteen!"

By then the Chinook, responding to the smoke, had come in and was approaching the downed bird. Fortunately, our hook man was already on top of the helicopter to be recovered, waiting to snap the dangling hook in place. The guide was leading the helicopter in.

I was on my back, aware of all this by sound only. I couldn't see a thing but felt water being poured onto my face.

"Chief, open your eyes! I've got to get water in your eyes!"

I didn't recognize the voice but learned later that he was a medic with the security platoon. I tried to open my eyes and may have even gotten them open, but the only thing I was aware of was the pain and that I couldn't see. Had I been blinded? Would I be blind for life?

I heard the Chinook departing and muttered, "How was the lift?"

"The lift was clean, Dick, don't be worrying about that," CW3 Bivens said. "Come on, we've got to get you back."

"Can you walk, Chief?" McKay asked.

"Yeah, I can walk, I just won't know where I'm going."

With someone on each arm, I made it back to the helicopter. They helped me in and lay me down on the floor in back.

"Keep pouring water in his eyes," the medic instructed.

I wasn't wearing my APH-5 (flight helmet), so I heard none of the conversation on the way back to Tan Son Nhat. The inability to hear anything but the whine of the transmission and the pop of the rotor blades left me feeling very isolated . . . and very afraid. I lay there for the 30 minutes of the flight back, alternating between praying that I wouldn't be blind and that I would be able to live with it if I never recovered my sight.

By the time we landed, the constant washing had begun to take effect. I could see Creech leaning over me, and while I couldn't yet see him well enough to determine whether he was looking worried or not, I felt surge of joy that I could see at all.

An ambulance met us as soon as we set down, and I was taken to the 3rd Field Hospital. The doctors and nurses of the 3rd Field shared the mess with us at the Red Bull Officers' Club. Maybe one of the nurses would give me special treatment, I secretly hoped.

I didn't see any of them. A medic continued to wash my eyes until, later that night, I was fine except for a slight burning sensation. I could see as well as I ever could.

The next day, I was returned to duty. This time my prayer was one of thanks.

Small World

Once, while stationed at Tan Son Nhut, I saw a young black soldier walking down the long stretch of empty road between the airfield and the barracks. Stopping the jeep, I offered him a ride, and he smiled and hopped in. I noticed the word Missouri on the side of his hat.

"You're from Missouri?" I asked as I pulled back onto the road.

"Yes, sir."

"Where in Missouri?"

"Oh, it's a town you've probably never heard of. It's called Sikeston. I sure wish I was there now. They've got a place there that sells the best barbecue in the world."

"Would that be Vaughan's Barbecue?" I asked.

The young soldier was clearly shocked. "Yes, sir. You mean you've heard of Sikeston and Vaughan's Barbecue?"

I turned so he could read my name tag. "I sure have. My dad owns Vaughan's Barbecue."

We conversed for the rest of the short trip until I dropped him off in front of his barracks.

Several years later, when I was publishing the newspaper *Delta Metro*, I wrote about this incident in an editorial. The next day, I was driving by the post office when I heard someone shout: "Mr. Vaughan!"

Looking toward the shout, I saw the very same man smiling at me from the sidewalk.

"It's me!" he said, throwing his arms out. I knew immediately what he was talking about.

We went to Vaughan's Barbecue, and I talked Dad into giving him a side of ribs. Which was fine . . . but I couldn't talk my dad into giving me any ribs.

The Crawling Peril and the Steak Sandwich

During my second tour in Vietnam, I was with the 56th Trans Company. The officers were billeted in the Red Bull Inn, the name we gave our BOQ (bachelor officer quarters). The Red Bull consisted of a series of interconnected cabins—very much like a motel—and an attached Officers' Open Mess.

We were billeted two to a room, and my roommate was CW2 John Usher. One evening, after normal duty hours, John and I returned from the airfield and were going to stop by our room to wash up for supper. But, when we opened the door, *the floor was literally black with roaches!*

You know how some people are repelled by spiders and some by snakes? For me it is a roach. I don't know why, but I can't stand to even look at one. And this is from someone who has lived nearly half a lifetime in Alabama. Even today, if I see a roach, I have to call Ruth to take care of it. Over the many years we've been married, she has come to accept this and will even warn me not to come into a room because she has seen a roach.

But back to the Red Bull Inn, and the floor that was black with roaches. When I saw them, I got dizzy and didn't know if I was going to throw up or pass out. I stepped back outside and braced my hand on the wall.

"Damn, look at that," John said, as if that disgusting display was merely an item of interest. "We'd better get the insect spray and go to work."

"John, you do it," I pleaded, barely able to speak. "Please . . . you do it. I can't go in there. I won't go in there until every roach is gone."

"Don't tell me you're scared of roaches," John said, surprised by my reaction. Then he saw my expression. "Damn, you really are, aren't you?"

"Please," I said. "Take care of them."

"Is it worth a steak sandwich if I do it?"

"Yes, yes, anything, please . . . just get rid of them!"

I went to the club to wait for him. It had been three years since my last drink of alcohol, but I nearly broke my pledge that day. Contenting myself with a Coke, I held a table until he showed up.

"All right they're all gone," he said about half an hour later. "It'll be a couple of hours before we can breathe in there, though."

"That's all right. Thanks, John, I'll never forget this." (As an aside, it is now 50 years later, and as you can see, I still haven't forgotten.)

"You said something about a steak sandwich?"

"Yes, yes, order away. It's on me."

When the Vietnamese waitress came to our table, John ordered his sandwich. "This is how I want it made," he said. "Instead of two pieces of bread, I want two steaks and one piece of bread."

"What? You *dinky dau!*" Miss Kim replied with a laugh. "That not one steak sandwich, that two."

"It's okay, Vaughan is paying for it," John said.

"You pay?" Miss Kim asked.

"Yes."

"Now I think you *dinky dau*."

"Ha!" John said. "If you think he's crazy now, you should have seen him half an hour ago with all the—"

"John, don't even talk about them," I pleaded.

John shook his head. "Miss Kim is right. You are *dinky dau*."

The Soldiers' Library

While I was in-country, I began writing articles, called *Vietnam Montage*, which were published in a few newspapers back in the States. I wrote one requesting that people send servicemen books instead of Kool-Aid. My 16-year-old cousin Sheila Kay decided to make that one of her high school projects, and within a month after my article appeared in the *Jackson Clarion Ledger*, as many as 20,000 books arrived in-country on a Mississippi Air National Guard C-130, dedicated to that very purpose.

I had intended for it to be maybe a hundred or so for the company library, but with so many books, they were distributed throughout the entire US military in Vietnam. In gratitude, we named our recovery helicopter the *Sheila Kay*.

Two weeks after the *Sheila Kay* was christened, its crew attended a briefing. The First Infantry Division was about to insert troops in an area near the Michelin Plantation (yes, Michelin tires). We weren't a part of the First ID, but we did support them.

The briefing pointed out the landing zones where we

would be making our insertion, along with a best-guess estimate as to the expected ground fire. We reviewed how quickly each wave should go in, off-load its troops, then take off, with the next wave literally landing in the rotor wash of the previous one.

We would assemble on the Michelin Plantation's private airstrip, where we would receive the latest intel on the landing zones and the amount of resistance we could expect. I was riding in the left seat, which meant I was the co-pilot to Chief Warrant Officer 3 Bob Bivens, who was from Caruthersville, Missouri, just 50 miles south of where I was from. Sergeant John McKay, a young black soldier from California, was our crew chief, and SP/4 Mike Pounders, who was from New Orleans, was our door gunner.

There was a surreal quality as we made our approach over the plantation. At least half a dozen "round-eye women"—our slang for the French—were lying in their bikinis beside the private pool. Two Beechcraft Bonanzas were parked on the airstrip, and at least 40 Hueys, who had preceded us, formed one long row along the airstrip. We could see soldiers spending their last few moments before the lift walking around, talking, and joking with each other to ease the tension.

Bob made his approach and flared out less than 20 feet behind the last helicopter in the queue. He came to a controlled hover about three feet above the ground, then eased the collective down so we could complete our landing.

As soon as we touched down, there was a huge explosion! The *Sheila Kay* shook, and the main rotor blades dug into the ground.

"What the hell was that?" I shouted. I must confess that it was more an expression of terror than a question seeking an answer.

Pieces of the tail cone came fluttering down all around us.

"I landed on a mine!" Bob said, repeating as if in a trance, "I landed on a mine! I landed on a mine!"

"Chief, the ship's on fire!" McKay called.

"I landed on a mine!" Bob shouted again.

"Bob, we've got to get out of here! We're on fire!"

That was enough to snap Bob out of it. He opened the door, pulled back the chicken plate (armor beside the pilot's and co-pilot's seats, designed to turn away anything shot at you), jumped out, and ran from the ship. McKay, who was on my side, and Pounders on the right side jumped out as well. Only I remained. I tried to slide my chicken plate back but couldn't make it move.

"Get out of there, Chief!" McKay shouted.

I could feel the heat of the fire, but try as I may, I couldn't make that chicken plate slide out of the way. I was trapped. Then I saw McKay running toward the burning helicopter. He stuck his arms in to grab mine, and with him pulling and me pushing with my legs, I was able to escape the ship *over the top* of the chicken plate.

There was no secondary explosion, but we barely got away from the *Sheila Kay* before the rounds started cooking off, sounding like very loud bursting popcorn kernels.

By now there were 40 or 50 of the air assault troopers gathered to watch the helicopter burn. "What happened?" one of them asked.

Bob replied, much more calmly this time: "I landed on a mine."

No Knock Warrant

I applied for and was given the opportunity to "live on the economy" in Saigon, so I found an apartment on Truang Dhogn, a street better known to the GIs as Plantation Road. It was very close to an establishment called The Pretty Girl and Happy GI Bar. It was perfect for me: I had a staff car, which I had on an open-hand receipt from the US Navy Saigon motor pool, even though I was from the army, and there was a secure US military parking area within two blocks of my apartment.

I liked living in the apartment. I was writing my picaresque novel, *Brandywine's War, Back in Country*, and living in my own, private apartment, which provided a lot of quality, alone time. I had all the comforts of home, including a small refrigerator that I had bought at the PX, a cooking stove, a table, a desk, a sofa, and a comfortable bed. Early every morning I would be awakened by the sound of soup boys clacking their sticks together to advertise the strolling soup vendors. *Mamasans* were always nearby, singing, "Bunmae! Bunmae!"—a very delicious baguette. The rhythm of the clacking sticks was unique to a specific soup vendor, and I would listen for the one I liked, then meet him in the street with my bowl.

I also had a typewriter and paper, so what more could I possibly want? The typewriter was a portable electric Smith Corona. I had grown used to using an electric typewriter, but there were times I regretted it, because Saigon was subject to brown outs—even total black outs—and when that happened, I would have to sit in the total dark, unable to write or even read.

Fortunately the brown/blackouts weren't too frequent, so I was able to get a lot of work done. I was in constant communication, by mail, with my editor in New York. Unlike now, when email can afford an instantaneous exchange of manuscript and revisions, these exchanges were, at a minimum, two weeks apart. Still, work on the book progressed.

"I know there's a war going on, not only for you, but for your character, but don't you think you could give Brandywine some sort of love interest?"

"Part of the thing that's driving this novel is Brandywine's determination to be somewhat more insane than the war itself. That's the only way he feels he can survive."

"Yes, but can't he be a little insane with some female? A Vietnamese woman? Or maybe an American nurse or something?"

"All right, I'll introduce Lady Jane Grey."

"Wait, wasn't Lady Jane Grey beheaded?"

"This one won't be."

This exchange with my editor took almost three months, but it introduced Jane Gray, the nurse in *Brandywine's War*.

As it turned out, though, my little quiet getaway had a drawback. A lot of American soldiers were using their off-duty time to rent apartments for activities other than writing a book. And whereas I had off-base quarters authorization from the base billeting command, many did not. And the nefarious opportunity most practiced by these "shadow soldiers" was the illicit use of drugs. From time to time, US military authorities would make a sweep through the area in pursuit of these soldiers, some of whom had deserted the army and were now supporting themselves by black market and drug dealing.

One night, I had written until late and was asleep very soon after I turned in. I was awakened with a loud bang, and when I opened my eyes I was blinded by a bright light and couldn't see a thing. In my grogginess, I thought that perhaps a VC sapper squad had broken in, and I reached for the pistol hanging from the head of my bed. The upstroke of a rifle butt knocked the pistol from my hand.

"Freeze!" someone shouted in English.

In the street out front, I could hear the popping sound and radio call of a PRC-10.

"Thank god you're Amer—" that was as far as I got before the barrel of a rifle was rammed into my mouth, preventing me from saying a word.

"Keep your mouth shut, soldier! Who are you?"

"How am I going to tell you who I am if I have to keep my mouth shut?" I tried to ask, but it came out "Hgg, mug, tgg, mug. . . ."

"Take the rifle out out of his mouth, Carter."

The rifle barrel was withdrawn.

"I'm Chief Warrant Officer Vaughan, and I have an authorized billeting assignment for these quarters."

"Let me see it, and some ID," the lieutenant said.

I showed him the documents.

"All right, men, he's authorized to be here. We've got other places to go."

They left without an apology and with the door still lying on the floor. The next day, I moved back into the Red Bull.

Assignment Manipulation

One of the first things you learn when you go into the army is that it's a very structured organization. A TO&E (table of organization and equipment) tells each unit exactly what they must have in the way of manpower and equipment. And it has a very organized rank system, so that everyone knows exactly what position they occupy, and unit structure, from squad to platoon, company, battalion, regiment, group, division, corps, and army (as in a specific army, such as 3rd Army, as opposed to the overall U.S. Army).

One thing about army structure is that, almost by definition, it is inflexible, and if you are in the army long enough, you realize that very inflexibility provides room to manipulate the system. I'll give you an example.

When I deployed to Vietnam for the third time, I was in the 34th Group, awaiting further assignment. I had been "up country" before and didn't particularly want to do that again. I knew that within two months the chief of the 110th Open Storage Depot would be returning to the States, and I wanted that slot. But I would have to keep myself in Saigon and in a position where I could be assigned there when the position came open, so I began the "manipulation" necessary to bring that about.

I went to the commanding officer of the 34th with a suggestion. "Sir, if you had an officer who was in charge of monitoring the readiness status of every aircraft for which we are responsible, it would give you an immediate picture of where we stand with regard to flyable aircraft, aircraft on Red X for repair, ADP aircraft awaiting parts, aircraft that are due intermediate and periodic inspections, and aircraft lost due to accident or enemy action. I could set up a status board and require reports be sent to me from all field units."

"Excellent suggestion, Mr. Vaughan!" the CO said. "How many men will you need?"

"Oh, one sergeant and a clerk is all. And an office."

"Consider it done."

Two weeks after the Aircraft Readiness Section was formed, the colonel who had authorized its organization rotated back to the States, but not before he awarded me an Army Commendation Medal for "establishing a system of monitoring readiness status of all aircraft in the field."

Six weeks later, the position I really wanted, chief of the 110th Open Storage depot, a command position that was rare for a warrant officer, opened up. I went to the new commanding officer of the 34th Group and asked to be given the assignment.

"I can't do that, Mr. Vaughan. I don't have anyone available to take charge of the Aircraft Readiness Monitoring Section."

"Sir, if I show you how that position could be eliminated, saving manpower and streamlining the operation, would you give me the Open Depot?"

"Sure, if you can show me how to eliminate it."

The truth was, what I was doing was completely superfluous, as every field unit was maintaining their own readiness monitoring and requesting, as needed, repair parts or replacement aircraft. There was no need for any oversight. I presented this report to the colonel.

"Excellent suggestion, Mr. Vaughan!" the colonel said after reading my report. "I am always in favor of streamlining the operation. How soon can you close up operations here and take over the depot?"

"I can do it within a week," I said.

"Good. Get right on it."

Within a week, I was in charge of the 110th Open Storage Depot, a much more challenging and personally rewarding assignment. I took my sergeant and my clerk with me, both of them happy not to be reassigned up country.

One month later, I received an oak leaf cluster to the commendation medal I had received for creating the Aircraft Readiness Monitoring section. The oak leaf cluster was for finding a way to eliminate that very section.

Personnel Adjustment

My third and final assignment in Vietnam was as commanding officer of the 110th Open Storage. Actually, because I was a warrant officer, I couldn't technically command, so I was "chief of."

In addition to the 40 American soldiers in my command, I had 10 Vietnamese civilians, plus two American civilians who worked, not for the army, but for RMK.

One of those civilians was Mr. Cupp, who was a problem. Half the time he would show up for work so zoned out that he just stood around in a daze. And actually it was better when he didn't do anything, because when he did, he would screw something up.

"Chief, Cupp just ran the forklift into the side of a crate of rotor blades and tore the hell out of two of them," Sergeant Murphy said.

"That's it, I've had with Cupp. Get him in here; I'm going to fire him."

Murphy shook his head. "You can't fire him. Captain Chambers tried. He doesn't work for the army; he works for RMK."

"Get him in here anyway."

A moment later, a very belligerent Cupp came into the office. "What do you want, Vaughan?"

"You're fired," I said.

Cupp laughed. "You can't fire me."

"Miss Nguyen, call the RMK supervisor for me," I said to my secretary.

A moment later, she had him on the line. "Cupp, you pick up the phone on Miss Nguyen's desk and listen in. I'll use the phone on my desk."

The RMK supervisor was a man named Glenn. "Mr. Glenn, this is Chief Warrant Officer Vaughan with the 110th. I thought I should tell you that Cupp just destroyed seventy thousand dollars worth of rotor blades. This isn't the first time he's done something like this, so I just fired him."

There was an audible sigh from Glenn, and he spoke to me as if I were some new person who didn't understand procedures. "I explained this to Captain Chambers, and now I will explain it to you. You can't fire him. He doesn't work for you; he works for me."

"I told the dumb son of a bitch that," Cupp said.

"Cupp's listening in on the other phone," I explained.

"Very good. So now you understand."

"You're right, Cupp doesn't work for me," I said. "But the Open Storage is a closed perimeter, and the only way in is by the front gate. The guards at the front gate *do* work for me, and I intend to give them instructions that if Cupp attempts to come through that gate, they are to shoot the son of a bitch. So, if you want to pay him to spend his day in some of the Saigon bars, be my guest. But he will not be allowed through the gate."

"Cupp, are you still on the phone?" Glenn asked.

"Yes, sir, I'm glad you got this cleared up."

"You're fired," Glenn said.

A Sad Memory

My TO&E vehicle was a three-quarter-ton truck.

"Chief," Specialist Jimmy Winston said. "You're the head honcho here. You ought to have a jeep." Winston had a personal interest in this, because he was assigned as my driver.

"Oh, I agree with you, Winston, but the TO&E doesn't."

"I can get you a jeep."

"Oh, you can, can you? How?"

"We can buy one from the Vietnamese for 500 sheets of plywood. And Lord knows we have enough of that."

"All right, do what you can do," I said.

The next day, two white jeeps, stopped in front of my office. They were Saigon police jeeps, which were white, matching the uniforms of the Saigon police, and the GIs referred to them as White Mice.

"Here's your jeep, sir," Winston said. "I've already delivered the plywood."

"Winston, I can't drive a white jeep around."

Winston smiled. "No problem, Chief. It won't by white long."

When I looked outside, I saw half-a-dozen of the Vietnamese workers painting one of the jeeps with spray cans. When that was done, Winston had already cut a stencil, and he held it over the bumper: *110 Trans 365th -3.*

"What do you think?" Winston asked with a proud smile.

"Looks fine to me," I replied.

"Now I won't be embarrassed driving my chief around in a three-quarter."

Over the next few months, the jeep came in very handy. And as we drove to Phu Loi or Di An or Long Bin, we developed a game. If we would see a flat-looking rock in the road, Winston would give me a shout out.

"There's a rock, sir! See how far you can drag it!"

Winston would steer so I could put my foot down on the rock and drag it along the pavement.

"Good, sir! That's a record!" Winston called out.

I would have believed him, but according to Winston, every dragged rock was a record.

One day we were on our way to Di An when an ARVN three-quarter truck came speeding around us, with the driver and a passenger in the front and four passengers in the back. We could see them because the canvas was gone.

"Holy crap," Winston said. "He's a fool to be driving that—"

That was as far as Winston got, because there was a huge explosion just ahead, and the burning truck flipped over, spilling the men out.

Winston braked hard, then we hopped out to see if we could help—but it was impossible to help pieces of bodies.

When the MPs arrived, we told them what happened, then continued on our journey.

"You know what, Chief? That could have been us," Winston said, voicing my own thoughts.

A few days later, Jimmy Winston came up to my desk. "Want me to take you to lunch, sir?"

"Well, I wasn't going to go just yet, but I guess I could."

We left the gate, but we drove passed the Red Bull, which was the officer's mess.

"Where are we going?"

Winston smiled. "A 'brother' retired out of the army and opened a soul food restaurant. I'm going to take you to have some soul food. Now, you might not be able to eat it—I mean, it's a brother thing—but if you can't, we can go to the PX and get you a hamburger or something."

When we stepped into the restaurant, I was assailed by delicious and familiar aromas. When the waitress came, I ordered breaded pork chops, black-eyed peas, fried okra, and turnip greens.

Winston perused the menu with a look of confusion. "I'll, uh, I'll have the chitterlings."

"Jimmy have you ever had chitlins before?"

"No, sir, but it's a brother thing."

When they brought the chitterlings to the table, with the accompanying pungent smell of hog intestines, Winston held his nose and waved the plate away. "No, no, get that out of here."

Winston watched me enjoy my meal, including some of his chitterlings, then we went to the PX for him to have a hamburger.

"How could you eat that?" he asked.

"Winston, it might be soul food to you, but to me it's home cooking."

Three weeks later, as SP/5 Jimmy Winston was taking some of the civilian work corps back home, the three-quarter truck he was driving hit a mine and, I'm sorry to say, Jimmy Winston was killed. I think of him still.

PTSD

We didn't have PTSD when we came back from Vietnam. Well, that's not entirely true. We did have it, we just didn't know what to call it. Part of it was the way we came home. Most of us flew home, and that was great—it would get us back with our loved ones in less than two days. That was also bad.

The soldiers returning from WWII and Korea came by ship, which gave them time for decompression, so the immediacy of the war was put behind them. Not so with the soldiers from Vietnam. We were home before we even changed clothes.

To be honest, it wasn't all that bad for those of us who were career soldiers. We came back to a CONUS base where nearly everyone had been in Vietnam or were going, which meant we had shared experiences. But the experience was quite different for the part-time soldiers, draftees, or those who enlisted for one tour and then left the army.

On a Friday morning, such a soldier might be on ambush patrol outside the fence at Cu Chi, Di An, or Phu Loi. When he comes back that afternoon, he has his travel orders, and the next day he boards a Pan Am flight for San Francisco. By Monday afternoon, he's sitting in the Bulldog Drive-in in Sikeston, Missouri, or the Purity Café in Greenville, Illinois, or some "burger doodle" somewhere in the U.S.

He is listening to the juke box and visiting with people he has known his entire life. Physically he is home, but mentally and emotionally, he is still back in-country. He remembers that Creech owes him $5. McKay has his sunglasses. He wonders if anyone will find the three cans of fruit from the Cs that he hid behind the 500 kW generator. Will the rod-end bearings that are on ADP orders come in tomorrow? What about the tire for the three-quarter-ton truck? He looks around at the others—they are laughing, teasing, talking about things that were once so important to him. The football game with Poplar Bluff was two weeks ago, someone comments, and he remembers that was the day Lambdin was killed. He realizes that, though he had grown up with these people, they were not there for a very important part of his life. None of them have any idea—nor do they care—where he has been, what he has done, or what he has seen.

Where is Schuler? Where is Kirby? Chambers? Lindell? They're still back in-country. Winston is still there, as well, but like Bostic, Lambdin, Morris, Wyatt, and Karnes, he will be there forever. Oh, their bodies are back, but they didn't make it. Bivins is back, though, somewhere in Ohio. Is he going through the same thing?

The one-tour draftee stares at Lucy, a girl he had dated a few times, and she is smiling at him. But her smile is replaced with a sudden flash of fear, and he quickly glances away. He realizes that he has given her the *thousand-yard stare*, and he should apologize, but he can't. She would never understand, and he's not sure that he does.

It's been 50 years, but even now, a song, a smell, a sight, will bring it all back... tone and tint. And if we see someone wearing a Vietnam Vet's cap, we'll nod and say something like, "Welcome home, brother." Others seeing us will see two old men . . . but they don't see what we see. We are greeting a young man in jungle fatigues, maybe standing on the service deck of a Huey with the engine cowl removed, or wearing a flak jacket and carrying an M-16, or leaning against a jeep with his arms folded, or sitting on a sandbag-reinforced Conex container, writing or reading a letter.

I often wear my Vietnam Veteran cap as I walk Charley, and from time to time someone will thank me for my service. I appreciate that, and I accept the thanks on behalf all who served in Vietnam, two-thirds of whom are gone now. The Vietnam War was such a divisive part of our history that many of the 1.6 million Vietnam Veterans who have died were never thanked, and rarely even had their service acknowledged.

Earlier, I told what it was like to go to Vietnam. Now let me tell you what it was like to come home. I have heard it said that the "spitting on the returning soldiers" is a myth. I was never spat upon, but I'm sure some were. The point is, "spitting on the returning soldiers" has become a metaphor for what did happen. And the longer the war lasted, the more hostile the protests became. I remember returning from my third deployment, getting on the army bus at OAT that was to take us to the airport. There was a very heavy wire mesh across the windows, something that hadn't been there before, and I asked the driver why it was there.

"You'll see," he said.

As we drove through the gate, there were perhaps a dozen or so people gathered outside. They began throwing rocks, bricks, big dirt clods, and bottles at the bus. The rocks and bricks just bounced off, but the dirt clods and some glass shards from the bottles managed to make it inside. As I waited in San Francisco Airport, a young boy who was perhaps six or seven came up to me, intrigued by my uniform.

"Are you in the army?"

"I sure am."

"Billy, you get away from him now!" his mother said angrily.

"Oh, he's not bothering me, ma'am."

"I don't care whether he's bothering you. I don't want him talking to a baby killer."

"Ma'am, you have the wrong idea," I said. "If the babies don't dress out to at least forty pounds, we generally threw them back."

She glared and cursed at me, evidently not understanding sarcasm.

Too many of us also returned to broken marriages, and I was one of them. The tale of my homecoming from my last tour in Vietnam is short but not so sweet. Gail and I had been married for six years, and I spent three of those years in my three tours in Vietnam. I called Gail from San Francisco with the happy news that I was back in the States and she could pick me up at the airport the next morning.

"I sleep late on Saturdays, take a cab," she said in a voice devoid of any emotion. "And don't wake me up when you come in."

Homecoming

I was hurt and confused during the long flight back to Norfolk, Virginia. What was going on? This had been my third tour to Vietnam, and I knew it would be my last. There would be no more long separations, I wouldn't be seeing green tracer rounds coming toward me. This should be a happy thing for both of us. I was home. But, what would happen now?

I passed through Chicago on my way from Oakland Army Terminal to my next duty station at Fort Eustis, Virginia.

Four months earlier, my driver, SP/5 Jimmy Winston, had been taking some indigenous employees to Di An when the three-quarter-ton he was driving hit a road mine and he was killed.

I had Jimmy's telephone number, so I called his mother. She already knew about it, of course. Jimmy had come home in a flag-draped coffin four months earlier. I hesitated for a second, just before I dialed. What if all my call did was bring back that awful memory and she started crying? How would I handle that?

Ring... "Hello?"

"Mrs. Winston?"

"Yes."

"Mrs. Winston, my name is Chief Warrant Officer Vaughan and—"

"Oh! You were Jimmy's officer!"

"Yes, ma'am."

"Jimmy talked about you so much in his letters. He said he wished all the officers he had worked for were like Chief Vaughan."

This wasn't going as I had planned. I intended to give her some words of comfort, but her words were making me choke up.

"I'm just passing through Chicago and I wanted to call you, to tell you what a wonderful young man Jimmy was and how much everyone in the company liked him."

"That is so sweet of you. And thank you for the letter you sent."

"I'm just sorry I had to send it."

"Jimmy wasn't drafted. you know. He joined," Mrs. Winston said proudly.

"Yes, I know. He was very proud that his serial number started with RA."

There was a long period of heavy silence before she spoke again. "Chief Vaughan? I know making this call was hard for you. Have you had to make many?"

"One is too many."

"Bless you, Chief. I know God has a special place for you."

By now my eyes were filled with tears. "Uh, Mrs. Winston, they're calling my flight."

They weren't, but I just didn't want to break down over the phone.

"All right. Thank you so much for calling. I'm going to share it with the whole family, and next time I visit Jimmy's grave, I'm going to tell him that his officer called. I know he will be so proud."

I hung up the phone and started back toward the gate.

"Mama, that man's crying," a little girl said. "He's a big man, and he's crying."

"Hush, sweetheart. Sometimes people just hurt." She nodded sympathetically at me, which was not the way we were often treated when we were in uniform, in those days.

Wondering what kind of reception I'd get when I got home, I stepped into the house and my dog Sheri jumped up on me in absolute joy. I leaned over, and she showered my face with doggie kisses. I was sure Gail would kiss me, and that would take away the hurt from the way she sounded on the phone.

I hadn't eaten anything since the noon meal yesterday, so I thought I would cook myself some bacon and eggs. There was no bacon or eggs in the refrigerator, but there was ground beef, so I began cooking a hamburger.

Gail came into the kitchen then. "You're stinking up the whole house," she said in a disgusted voice.

"I'm sorry. I was hungry."

"I'll open a window," she said.

"You want me to fix one for you?"

"No."

I ate the hamburger, alone, then went into my office. One of the things I had thought about and really looked forward to while I was in-country was my book collection. I had 600 or 700, including autographed books by Pearl Buck, Norman Mailer, Joseph Heller, and several lesser-known writers who I had met in the brief writing career I had already established.

They were gone. Every one of them. The bookshelves were filled with little vases and knicknacks.

"Where are my books?" I shouted in absolute horror.

"I gave most of them away and burned the rest of them."

"Why?" The question came out almost as an agonized cry.

"They were beginning to stink. I want a divorce."

The Today Show . . . Almost

I moved out of the house and took an apartment in downtown Newport News. Within a month of my return, I got a contract for another book, and though I was still in the army, I was able to work on it after duty hours. One day I got a call from my publisher.

"You're going to be on the *Today* show," Sam Post said, the excitement in his voice coming through the telephone. "Wear your uniform with all the ribbons," he added.

I was at Fort Eustis, and the reason I was going on the *Today* show was to publicize my book, *Brandywine's War*, which had reached several best-seller lists. It was an iconoclastic look at helicopter operations in Vietnam, and *The New York Times* said that the book with its "dark humor" was to Vietnam what *M*A*S*H* had been to Korea and *Catch-22* had been to WWII.

This was in the day when you could show up at the airport and book a flight as cheaply one hour before takeoff as a month earlier. Also, first class from Newport News to New York cost only $12 more, so I took a flight the very next morning—first class. Hey, why not? I was going to be on the *Today* show.

I took a cab from JFK to the offices of MacFadden-Bartell, publisher of the book, and was greeted effusively by many of the people there. The book was climbing national best-seller lists, and that always helps a writer who shows up at a publishing house.

"We've booked a room for you at the Algonquin," Sam said. "That's only two doors down from the Harvard Club, and we'll have dinner there."

"Harvard Club? Sam, I had one, very unproductive semester at the Missouri School of Mines in Rolla."

"That's all right, I have a masters from Harvard."

The reason for dinner at the Harvard Club became obvious when I met our dinner guest: Ralph Daigh, president of Fawcett and Ernest Hemingway's last editor. The purpose of the dinner was to sell paperback rights to the book.

"Tell me about the book, Dick," he invited.

"It's about helicopters in Vietnam, but it's not a war story. It's more . . . well . . . there's a five hundred dollar rat, and a soldier who can pee over a fence and, uh . . ." I came to a complete halt, trying to describe the book.

Ralph took a book of matches from the ashtray on the table, tore the matches out, and handed me the cover with a ballpoint pen.

"If you can write, on this matchbook cover, what the book is about, I'll buy it."

I did, and he did.

When we returned to the Algonquin that night, Sam said, "Pearl Buck will be on the *Today* show with you tomorrow. We're publishing her *Story Bible*. NBC will send a limo for you tomorrow morning, so be ready."

"Hey, why don't you tell Pearl Buck she can ride in my limo."

Sam laughed. "You'll be riding in her limo. NBC would not have sent one just for you."

When the limo came to a stop in front of the Algonquin the next morning, the doorman hastened to open the door for me, which was pretty heady stuff for this Missouri boy. Pearl Buck was already in the car, and she greeted me. "Hello, Mr. Vaughan."

"Hello, Miss Pearl," I replied nervously. "Uh, I mean Miss Buck."

She laughed. "Sam said you were from the South, and I'm told that's quite a respectable term of address. What should I call you?" she asked, looking at my uniform. "Lieutenant? Captain?"

"Mister," I said. "I'm a warrant officer, and we are addressed as Mister."

When we reached the NBC building, we were shown to the Green Room. A moment later, Hugh Downs entered, greeted everyone, and turned to me.

"Mr. Vaughan, it has just been announced that President Nixon is going to visit China. That is a huge story, and given Miss Buck's familiarity with China, we have decided to give her the entire program. I'm sorry, we won't have time for you. But do enjoy the coffee and pastries."

A few years later, after I had left the army, I was living in Phoenix. My book *The Valkyrie Mandate* was out, and I was invited by a local Waldenbooks to do an autograph signing. Hugh Downs had since left the *Today* show and was now living in Fountain Hills, an upscale community that was very close to Phoenix. As it turns out, he had also been invited to sign one of his books. I suspected I was going to be shot down again.

"We've met before," I said and told him about the show with Pearl Buck.

"Yes, that was a great show," he said. He didn't remember that I had cooled my heels in the Green Room.

But things turned out well. *The Arizona Republic* had just done an article about me, and I had been on local TV promoting my book, so I sold a dump and a half . . . while

Hugh Downs was sitting there twiddling his thumbs. After selling only six books, he left.

I can't gloat too much, though. I and every writer I know have suffered through that dreaded "sit at the signing table and be ignored."

A Civilian at Last

My Television Career, Part I

After retiring from the army, I applied for and got a job with WAVY-TV. I was an on-air commentator, though I never actually did news features. I did what was called "soft features." But I did write news stories for the six and eleven o'clock news. This was, by the way, in the days of SOF (sound on film) rather than video.

One of my interviews was with a pretty young lady named Kaye who was a member of a professional theater group. The manager of the theater group left town with all the accounts receivable and had paid none of the accounts payable. That meant that Kaye and every other member of the production company were left high and dry. Our interview had been planned before this happened, and to be honest, I was a little surprised she kept it.

"What are you going to do now?" I asked.

"I don't know what I'll do."

"I'm about to spend a week in the Bahamas. Would you like to come with me?"

I had asked her as a *schtick* for the show and was surprised when she replied, "Yes, I would love to go."

We had been in Nassau for three days, and I was at the window of my room at the Sheraton Hotel, looking out over the harbor, where a three-masted sailing ship had arrived during the night. It was bulky and black and looked exactly like an old pirate ship. Fascinated, I went down to the hotel dock and rented a rowboat, then rowed out to the ship.

"Ahoy the ship!" I called, trying to sound very nautical.

A young man came to the railing and looked down. "What do you want?"

"Just to look around, if you don't mind."

"Don't mind at all; come aboard." He dropped a rope ladder over the side.

The ship was a 100-year-old Baltic Trader called the *Fremad*. There were five young people aboard—three men and two women—the oldest being 23. The three young men had bought the ship at auction in France. They didn't know each other beforehand but had all attended the boat auction and decided to put their money together to buy the *Fremad*.

It was their intention to sail around the world, but when they reached Nassau they were out of money, nearly out of food, and the maneuvering propeller had fallen off right there in the harbor. After a bit of discussion, I offered to restock their larder and to pay to have the propeller repaired, in exchange for them taking Kaye and me back to Fort Lauderdale, where I had left my car. They agreed, and the adventure was on.

One of the young women on board was a longtime girlfriend of one of the men. The other was an exceptionally beautiful young Frenchwoman named Chantal, and I was told she had lost her passport in a storm coming across. They asked if I would take her to the American Consulate and tell them that she worked for me as a translator of my books. To prove the point, I bought one of my books from a bookstore in Nassau and went to the consulate with her.

We had no problem getting the visa.

I had written about sailing ships, but this was my first experience on one, and it was great—as if I were reliving one of my own books. But as we came through the Gulf Stream, water started pouring in through the cracks between the strakes. Belowdecks we were ankle deep in water, and we had to keep the bilge pump going constantly—which unfortunately was a manual bilge pump. It turns out that they had not used pitch when they prepared the boat for sea but had used window caulking, which ultimately failed.

We made it through the stream, then limped into port. When the customs officials came aboard, they didn't question Chantal's newly issued visa. When they left, there was a collective sigh of relief from the others.

"What's wrong?" I asked.

"Chantal didn't lose her passport; she never had one. She came aboard during our *bon voyage* party, got drunk, and passed out belowdecks. We didn't even discover her until the next day out, but when we offered to take her back, she said she would just as soon stay with us. All she had with her was her bikini. She was on work release from jail . . . for passing bad checks."

"Oh," I said, feeling a little used for helping her get the visa. "What's going to happen to her now?"

"I don't know," the elected captain of the boat said.

"Well, you brought her here. Don't you feel some sense of obligation?"

"No."

"Chantal, do you have any money at all?" I asked, and she shook her head. I gave her a $100 bill, which she folded and put behind one of the cups of her bikini top. "What are you going to do?" I asked.

"Where are the expensive hotels?" she asked.

"You're on the beach in Fort Lauderdale. There are no cheap hotels."

She smiled. "I will be fine," she said confidently.

The last time I saw Chantal, she was walking down the beach toward the strip of hotels. She threw a wave over her shoulder but didn't turn around. I've often wondered what became of her, but I have the feeling that she turned out as she said she would, just fine.

My Television Career, Part II

Kaye and I began "keeping company" as they say, and because she had been left without a source of income, I hired her as my secretary, though I didn't actually need one.

I went back to the TV station. I had a rather warped sense of humor then, and I sort of let it gain control over me. In one incident, golfer Orville Moody came to visit. I convinced him it would be good if he gave some golfing lessons, on film, to our weather girl (I'm not being sexist, that's what they were called then). We went to the golf course, where he showed her how to hold the club, address the ball, etc. She did everything so clumsily that he was about to give up. Then he said, "Go ahead and hit it, and we'll start from there."

What I hadn't told Mr. Moody was that our weather girl had been a runner up national collegiate champion in women's golf. She knocked the ball about 250 yards, straight down the middle of the fairway. Moody's face reflected his shock. "*Dayum!* I should teach!" We had to cut that comment from the SOF.

Another incident involved our anchor. He was an excellent attack reader, and because of that he refused to discuss any news story before he presented it. He was a bit . . . I'll just say pompous. So I wrote a story for him for the six o'clock news:

Portsmouth pickle packers posted prices pertaining principal pursuits primarily protesting policing procedures yesterday.

He began: "Portsmouth pickle perk . . . uh . . . Portsmouth packer picker . . . uh...Por per pro . . . uh . . . but more on this at eleven."

He wasn't happy with me or the news story, which I insisted was simply impactful alliteration.

But my own personal demise—and what got me fired from the TV studio—was, I admit, a dumb and juvenile stunt. Earlier in the day a police officer had been a guest on a children's program, and he was showing all the equipment a policeman carried. When he left, he left behind his handcuffs. I was sitting at the desk having just given a movie review, and it was time for the weather report. The weather girl was working with the map in chroma key. In her right hand was a wand, and she was pointing to a blank blue background. Her left hand was on the corner of the desk—right in front of me and right above the policeman's handcuffs.

I couldn't resist. I handcuffed her to the desk on live TV, and the problem was . . . the policeman had left the handcuffs but not the key. It took us an hour to get her free, all while the station manager was watching from home.

That was the last day of my short TV career.

Managing a Congressional Political Campaign

The telephone call was unexpected.

"Mr. Vaughan?"

"Yes."

"This is Tommy Downing."

"You mean Congressman Tommy Downing?" I asked, referring to the congressman from the First Congressional District of Virginia.

"Yes, I've been watching you on TV, and I'd like to make a proposal to you. Do you think you could drop by my house and speak with me?"

"Yes, sir!" I said, surprised, a little bit intimidated, but mostly interested in what this was about.

The congressman gave me his address, and when I arrived at his house, I was greeted not only by him but by a very overweight man in a wheelchair. I noticed that both legs had been amputated just below the knees.

"Congressman Downing, I—"

"Please, call me Tommy. And may I call you Bob?"

"Well, yes, sir, though my family and all my friends

call me Dick."

Tommy chuckled. "I won't even ask how that came about, but if you want to be called Dick, then that's how it will be." He gestured toward the man in the wheelchair. "This is Tiny Hallman. I'm sure I don't have to explain the name Tiny. He's my AA."

"AA?"

"Administrative assistant. And as such, he's in charge of my reelection campaign. That's why I contacted you. I'd like you to manage my campaign."

"What? Oh, Congressman Downing, I'm not equipped to do anything like that. I mean, that's major stuff, and I have no political background at all."

"I've seen you conduct interviews on TV, Dick, and I know you can do it."

"If you've seen me on TV, you might also know that I was fired for handcuffing Rhonda Glenn to the desk during a live broadcast."

Tommy laughed. "I wondered why I didn't see you anymore. I remember that. I thought it was funny."

"Rhonda and the station manager didn't think it was funny."

"Let's get back to your managing my campaign. It will be a very easy campaign; I have no opponent."

"Really? Then I don't understand why you need a campaign manager."

"I still have to campaign, but mostly it will be touring around the district, speaking before various groups. All you have to do is coordinate those appearances. You don't even have to set them up—we've already got a lot of requests. You just have to schedule them. Oh, and drive the Winnebago. Normally Tommy would do this, but he just had both his legs amputated from diabetes, so I need someone to take his place until he can get back on his feet."

Tiny laughed.

"Sorry, Tiny. But I meant that figuratively, of course."

I agreed to manage the congressman's campaign, and even though he was running unopposed, the next two months were exhausting. We went from six in the morning until midnight, and I quickly learned the reason for the Winnebago. Tommy would sleep as I was driving between the cities, and often I would sleep while he was meeting with small groups. Anytime he spoke before a large group, I was there with him, soaking in the atmosphere of a major political campaign.

Because he had no competition, it was a foregone conclusion that Tommy would win, but we had a victory celebration anyway, and Tommy introduced me as the man who had managed his successful campaign.

After Tommy was elected, he asked if I would mind working in his office for a few weeks. "I'm sure you have to do a lot of research for the books you write. I've got a couple of projects I'd like you to research for me." He smiled. "And, you're writing a book about the assassination of Diem. You might find some information you could use."

"Yes, thank you, I will," I agreed heartily.

The Valkyrie Mandate

On November 2, 1963, President Ngo Dinh Diem was assassinated in South Vietnam. The assassination was part of a coup perpetrated by Duong Van Minh, a general popularly known as Big Minh.

Three years later to the day, I was riding with Nguyen Mot, a Vietnamese civilian who worked for me. We were stopped at railroad crossing number 6 on Tru Minh Ky, waiting for the train to pass.

"This is where it happened," Mot said. The comment seemed to come from nowhere.

"This is where what happened?"

"This is where President Diem was killed."

"You mean here, at this crossing?"

"Yes." Mot reached into his back pocket, took out his billfold, and extracted a small piece of cloth about one square inch, covered with brownish looking stains. "This is his blood."

"What? How in the world did you get Diem's blood?"

"I was in the APC when he and his brother were killed,"

Mot said. Mot had been a soldier attached to the detail that was sent to the St. Francis Xavier Catholic Church in Cholon to arrest Diem and his brother, Ngo Dinh Nhu.

"President Diem had called Big Minh to surrender, so we were sent to pick them up," Mot explained. "We were told we would be taking them to Tan Son Nhut to be flown away. When we reached the church, they gave us no trouble, and they got into the MPC peacefully. We started back, but when we reached this place"—he waved his hand toward the train that was still crossing—"Tru Minh Ky, Major Nghia, and Captain Nhung jumped down with their bayonets and began stabbing and . . ." unable to come up with the English word, Mot made the motion with his hand to simulate hacking, "the two brothers."

Mot told me the story in gory detail, and because of his personal connection to it, I began to research what had happened. In addition to my conversation with Mr. Mot, I spoke with Father De Paul, the priest who had taken the last confessions of Diem and Nhu. I also read all the archival material in the *Saigon Post,* an English-language newspaper. The full names of the two officers who killed the brothers, reportedly under orders from Big Minh, were Major Duong Hieu Nghia and Captain Nguyen Van Nhung.

Shortly before Diem surrendered, he called Ambassador Henry Cabot Lodge Jr., saying that he was willing to step down, and asking Lodge for an American guarantee of safety for him and his brother. Lodge passed him off. Although it quickly became apparent that the United States supported the coup, we had expressed, though not too strenuously, our opposition to the brothers being killed.

When I returned to the States, I tried to arrange a meeting with Lucien Conien, the shadowy American figure who was the liaison between the CIA and Big Minh's coup element, but I was unable to speak with him. By then Madam Nhu, the beautiful Dragon Lady of Vietnam, was in Paris. I tried to interview her, as well, but she wanted $25,000, so I declined.

The result of all this was my novel *The Valkyrie Mandate*.

Publishers Weekly said of the book, "Not since *Casablanca* have exotic settings and international intrigue been so richly and powerfully combined." A cable that I had in the book was read into the Watergate hearings. Because of that, Henry Cabot Lodge called me a damnable liar. I assume *damnable* is an even greater insult than a simple *damn*.

One of the first things a writer learns is the difference between truth and fact. All facts are true . . . but not all truth is factual.

The Valkyrie Mandate was nominated for the Pulitzer Prize, but the Pulitzer committee decided that no work of fiction was worthy of the prize that year. I would rather have lost to someone.

Eventually I was visited by members of the FBI.

"Mr. Vaughan, you once worked in Congressman Downing's office, did you not?"

"Only for a few weeks."

"Is that where you got the cable? It's classified, you know. You had no authority to release it."

"It isn't classified . . . in fact, it isn't real. I made it up."

"Are you saying you fabricated that cable?"

"Yes."

"Why would you do such a thing, Mr. Vaughan?"

"Wait a minute, you are talking about the cable *The Valkyrie Mandate*, aren't you?"

"Yes."

"Then, why are you asking me such a question? That's a novel."

"But much of it is historically accurate."

"Yes, when you write a historical novel, you have to be authentic as much as possible. Diem and his brother were both killed, and his government was overthrown."

"But you lied about the cable."

"By definition, a novel is *fiction*!" I said as assertively as I could.

"This cable was read into one of the committee hearings. And you are saying that it isn't true?"

"Yes," I said.

"Wait here."

The two men stepped of my hearing for a moment, then returned. "Would you be willing to sign an affidavit declaring the cable to be false?"

"Yes."

"We may not need to go that far; we'll be in touch."

Thankfully, I never heard from them again.

Wounded Knee

"Why don't we go to Wounded Knee?" Kaye asked.

"What for?"

"Haven't you been following the news? The Indians are demonstrating there. You could probably get a book out of it. If we could get in touch with Russell Means, we could get it set up."

Russell Means had been in the news a lot, because he was the head of AIM, the American Indian Movement.

"We'll need to find out where he is," Kaye said.

"No problem. I still have friends in the news department at WAVY. I'll ask one of them."

Russell Means was in Cleveland, Ohio, where he was to speak in a couple of days. I called him.

"Mr. Means, I'm an independent journalist. I wonder if we could set up an interview?"

"Sure. I'm in room 814. Come on up."

"Uh, I'm not in Cleveland, but I'll be there by tomorrow."

Actually Kaye and I got there that night, and I dialed 814.

"Mr. Means, this is Robert Vaughan. I called earlier. I

managed to catch an early flight, and I'm down in the lobby."

"Come on up," he invited.

Kaye and I knocked on his door no more than a minute later, and he invited us in. We talked for a few minutes about what he was planning to accomplish with AIM, then he showed me his speech.

"I'll be giving this at Case Western tomorrow night. What do you think of it?"

I looked at it and winced. "Uh, I, uh, think it's fine," I said.

"No, you don't. You hate it, and I hate it too. I've got a portable typewriter here. How about you write one for me?"

The university sent a cab for Russell the next evening, and Kaye and I accompanied him to his speaking engagement. There were at least 500 in the auditorium, and I was pleased to see that the speech was well-received, with applause where I thought it should be.

Afterward came the questions. Most were benign, but one of them had an unwanted reaction.

"Mr. Means, don't you think the average white man is in accord with the goals of AIM?"

"No, because if you were, you would all get guns, come to the Knee, and join us in this fight against government oppression."

That got a huge applause.

After the speech, two men wearing dark suits approached Russell.

"Mr. Means, you are under arrest for advocating the overthrow of the U.S. government. Who wrote your speech?"

"He did," Russell said, pointing to me.

"What is your name, sir?" one of the FBI agents asked.

"Robert Vaughan."

"Mr. Vaughan, you, too, are under arrest for advocating the violent overthrow of our government. Both of you are coming down to the police station with us."

"Kaye, take a cab back to the hotel," I said.

I went down to the police station with them, and I was grilled for about an hour but was never placed in a cell. Finally, after reading and rereading the speech, they were convinced that I was innocent, and they let me go.

After returning to Newport News, I bought a brand new van that had a bed and cabinet built into it, and I wired it for a hotplate and a refrigerator. We had a party with some of Kaye's friends the night before we were to leave, and when I saw the van the next day, it was covered with graffiti, the kind that you can't wash off. I took it in stride, and we took off for Wounded Knee.

We spent an uneventful month there. I listened in on some of their meetings, and they said they wanted some input into whomever the Indian agent was to be. I thought that was a reasonable request.

They said they wanted South Dakota back. I thought that was an unreasonable request.

When we left Wounded Knee, we stopped at a motel and I called my agent to report in for the first time in a month.

"Where are you?" he asked.

"Oh, I'm in some little town, I don't know the name of it. . . ."

"Rushville," a voice said.

When I checked with the desk, they assured me that nobody there had been listening in on my call.

It was very cold in Rushville, and I was watching TV while Kaye was taking a shower.

"*The warmest place in the nation today is Phoenix, Arizona.*"

When Kaye came out, I told her we were going to Phoenix.

We were together in Phoenix for about six months. At the end of that time, I wanted to leave and she wanted to stay, so there was an amicable parting of the ways.

Wrapping Fish

When I left Phoenix, I returned to my hometown of Sikeston, Missouri, and established a residence there for the first time in about 25 years. At loose ends, I began working for my dad, who after years of trucking had gone into a new business and now owned a barbecue and fresh fish market. About two weeks after I started working for him, a lady came in to buy fish, which I cut up and began wrapping in butcher paper. Apparently, I wasn't doing that good of a job, because she told me I wasn't.

"You're not doing a very good job wrapping that fish."

"Well, damn, lady, you planning on sending this fish through the mail?"

She gave me a disgusted look, took her poorly wrapped fish, and left.

"I don't think I'll be needing you anymore," my dad said. He had overheard the exchange and fired me. His own son.

Third Time's the Charm

I was still without a new book contract, so I approached our local newspaper with an idea. I would produce a weekly Sunday magazine for the paper, concentrating on soft features. The publisher bought into the idea, and *Scope Magazine* was born.

For one of my articles, I rode all night in a police squad car. The policeman's name was Marshal, and I had gone to school with him. As we drove around, we passed a house that I recognized.

"Judy and Pat used to live here," I said.

"Pat still does," Marshal said.

"You're kidding me."

"Nope. She got divorced and came back home."

The next day, I knocked on the door of the house that I knew well from the time I had been dating Judy. Pat answered.

"Dickey Vaughan," she said. She hadn't seen me in a quarter of a century, and I confess to being flattered by the recognition. I asked her out again, and this time she accepted.

It didn't take more than two or three dates to realize we could have a friendly relationship but there was no spark for anything more.

Then she introduced me to Ruth, who played with her in a woman's volleyball league. I went to one of the games with my camera, as if I were doing a piece for *Scope Magazine*. I wasn't, there wasn't even any film in the camera. I just wanted to see Ruth.

After the game, I asked if they would like to join me for a cup of tea. The only restaurant I could remember was the War Drum, so that's where we went. Over tea, we exchanged some of our background. Ruth had taught school in Alaska, 200 miles above the Arctic Circle. She was divorced and the mother of two boys, who at the moment were visiting their father, a schoolteacher in Alaska.

We seemed to hit it off, so I asked if she would like to go out on a real date, and she accepted. As we were getting ready to leave after finishing our tea, I picked up the tab and suddenly realized I had forgotten to bring my billfold. I didn't have one cent with me.

With an understanding smile, Ruth paid the tab.

We went together for about six weeks, then I got a book contract. I was to go to the "fat farm" at Duke University and ghostwrite a book for Dr. Richard Stulke, the manager of the diet program. I asked Ruth to marry me, holding out as incentive the opportunity to have a honeymoon at the Duke University fat farm.

If you think about it, it seems like an inauspicious beginning, but it has lasted for 45 years. I'm just really positive that this marriage is going to take.

The Fat Farm

I had been hired to write *Thin for Life* for the director of the Duke University Diet. As promised, I took Ruth there on our honeymoon, while I attended the program as one of the patients. Only Dr. Stulke knew my real purpose there.

The program was run by the university, and we patients were required to board there, so we had private rooms in an on-campus facility. And, of course, we took our highly controlled, low-calorie meals in a dedicated cafeteria. One of the patients was the comedian, Buddy Hackett. He ultimately got upset with the strict program, and as a parting gift, he arranged for a large pizza to be delivered to each of the 55 patients in the program. Needless to say, after weeks of eating nothing but bees' knees and gnats' knuckles, discipline broke down and all the pizzas were eagerly devoured.

There was a psychologist on the staff, and he was constantly watching us, making observations in a little notebook he carried. During our meals, Dr. Musante would sit on a stool at the edge of the dining room, keeping an eye on our every movement. He didn't know that I was researching a book I would be writing for the director, and so I decided to play a little game with him. Any time I saw him looking my way, I would curl my arm around my plate, lower my head, and look away from him. Often he would reposition his stool, and when he did, I would change arms and look away in a different direction.

Whenever I met him in a hallway, he would say, very cheerfully, "Good morning, Mr. Vaughan!"

I would immediately look down and away, avoiding his direct gaze. "G-good mornin', Doc . . . uh . . . Doctor Musante." I would mumble the words, speaking so quietly that he could barely hear my response.

Dr. Stulke told me that Dr. Musante was making me a special case for his study. He was absolutely sure that my weight was a textbook example of being psychologically driven. Then one day he asked me to see him in his office. I debated whether or not I should tell him that "I'm not actually a patient, I'm just cleverly disguised as an overweight person, here to research a book."

That's what I should have told him, but as I was walking to his office, I happened to notice that there was hole in my jeans, just above my right knee, and all resolve to be good went out the window, just as it did when I ate the large pizza. I knocked on his door.

"Come in, Mr. Vaughan."

As I stepped into his office, I put my hand over the hole in my jeans. "Y-you wanted to . . . to see me, Doc . . . Dr. Musante?"

"Yes, come in and have a seat."

I approached his desk, walking like Chester in the old Gunsmoke TV series, which had the effect of accenting me holding my hand over the hole in my jeans.

"Mr. Vaughan, it has been my observation that your biggest problem is one of self-confidence," Musante said. "I think your overeating is a result of compensation for not feeling good about yourself."

"I . . . I feel good about myself," I mumbled.

"Why then, when I try to observe you in the cafeteria, do you look away?"

"I don't look away."

"And when we pass in the hallway or somewhere, you won't even meet my gaze. You look away, and you mumble your response. Speak out, man! Have some self-respect!"

"I . . . d . . . I don't mumble," I mumbled.

"You're mumbling right now. And another thing . . . why are you covering that hole in your jeans? Good heavens, man, do you think someone might criticize you because you have a hole in your jeans?"

"I . . . I don't have a hole in my jeans."

"Of course you do. You're covering it right now, with your right hand."

I moved my left hand over and slid it under my right so that it covered the hole, then I raised my right hand, palm toward him. "No, I'm not. See, here is my right hand."

Dr. Musante sighed and shook his head. "That's all, Mr. Vaughan."

"I . . . I can go now?"

"Yes."

I left, then once outside I looked back around the doorway. Dr. Musante was writing feverishly in his notebook. Okay, that was bad of me, I know, but I just couldn't resist.

Later, Dr. Musante wrote his own book. and I'm sure his "case study" of me provided some of the material.

Cape Cod and the Persian Carpets

Back when my publishing domain was mostly New York City, we used to take frequent trips there. We took the boys with us, and they always seemed to enjoy it and would sometimes sit at their own table in the restaurant while Ruth and I discussed business with my agent or editor. They loved doing that and often would order strawberry shortcake as a main course, followed by ice cream and a slice of apple pie.

On one trip, we had planned to go on to Cape Cod after we concluded our business in New York, but when we were out strolling around, Ruth saw a crystal serving bowl with a silver rim, and she just had to take a closer look. A half hour later, we left the shop with the crystal bowl *and* two Persian carpets, reduced in price because of the Iranian hostage crisis but still too expensive for us to continue our trip beyond New York. As I write this, Cape Cod lies peacefully on our living and dining room floors.

Misspent Career?

I sold my first book when I was 19 years old. That was 63 years and north of 400 books ago: 400+ books, seven of which made the NY Times best-seller list, two of which were number one on the list.

But nobody has ever heard of me. Why? Because I have written under 52 pseudonyms, several of them under names you most likely *have* heard of. But contractually I'm not able to tell you all the names.

I wrote romance novels for Patricia Matthews, and I remember when one of them—*Love's Sweet Agony*, a novel about the Kentucky Derby—was number one on both the *Publishers Weekly* and *The New York Times* best-seller lists. Patricia was Gov. John Y. Brown Jr.'s personal guest in his box at the Kentucky Derby.

"Here is Patricia Matthews, author of the number-one book in the nation, about this very event, enjoying a mint julep with the governor," the announcer said, and as she waved and smiled, the real author was sitting right here on my sofa, eating popcorn and drinking a Diet Pepsi.

Once I was in the St. Louis airport, and I saw a woman reading one of the Paula Fairman books I had written.

"Are you enjoying the book?" I asked.

"Oh, yes, very much. It's a wonderful book," she said.

"I'm happy to hear that. Would you like me to autograph it for you?"

"What?"

"The book you are reading. If you'd like, I'll autograph it for you."

She looked at the author's name, then at me, and possessively pulled the book to her chest. "I think not," she said rather sharply.

It has been a bittersweet career. I'll get a book on *The New York Times* list, and I can tell Ruth and my dog about it. I can go to a bookstore and see it posted in the best-seller slot and feel a sense of satisfaction that *I put the words in that book*. But I can't tell anyone else.

I don't feel cheated. In every case I have signed every contract with my eyes wide open. I have done this willingly, because I'm too old to go back into the army, and my professional football career consisted of a two-hour tryout earning me an orange, a tuna-fish sandwich, and a promise to call me if they could use me. That was over 60 years ago, and I'm about to give up hope that they'll ever call.

Now, thanks to Wolfpack Publishing, I am beginning to establish my own name, and I'm happy to say that if you go on Amazon and type my name in the search box, well over a hundred books with *my* name will appear.

The Vietnam Wall

A few years ago, I told Ruth I would like to visit the Vietnam Wall. I have several friends whose names are there, including two of the eight women whose names are on that wall.

I must say that I was unprepared for the impact it would have on me. I stood at the top of the incline, looking toward the long, gleaming black slab, the more than 58,000 names just a blur. People were moving slowly along The Wall, some too young to have any memory of the Vietnam War, but most with hair that had been bleached white by the passing of too many years. Near where I waited, a couple of men were staring at one of the names, engaged in a quiet conversation.

"I don't believe Smitty ever bought a cigarette of his own."

"No, but he was a good man. You needed somethin' done, you could always count on him."

"Problem with him, he couldn't keep his mouth shut. Ole' Smitty was always talkin' himself into trouble."

"He only had three weeks till his drop date when he was killed. Can you believe that? Three weeks."

Ruth had gone to check in the book to help me locate some of my friends. They aren't listed alphabetically but are on The Wall chronologically, according to when they were killed.

She came back with the names. The first was Dan Lambdin.

Dan and I were friends from Germany, and though we were in Vietnam at the same time, we were not serving together. I found his name on the wall, and as dramatically as if it were a scene change in a movie, The Wall and the people around me, tone and tint, disappeared. I was no longer in Washington, DC, but was standing on a flight line, and I could hear the sound of rotor blades, smell the jet exhaust, and feel the oppressive heat of being "in-country."

Dan was in Vung Tau, and I had flown over from Phu Loi to deliver a helicopter just out of maintenance. Dan's roommate was gone, so he invited me to share his BOQ room. We lay there that night, filling the dark space between our bunks with the conversation of close friends, laughing over recalled incidents from Germany, and sharing our most personal thoughts and concerns about being in Vietnam.

"The first thing you have to do is to learn not to be afraid you're going to be killed every time you go up," Dan said. "You need to stop worrying about your mortality and think only about the momentary reality."

But we wondered what it would be like to be killed. Are you aware beyond death? Where does your soul go?

"Ha! You better believe my soul won't be staying here," Dan said.

The next morning, we went down to the airfield together, me to take a replacement helicopter back to Phu Loi, Dan to deliver a generator to a Special Forces "A-Team" unit near Binh Khat. We exchanged some off-color remark by way of goodbye, then I heard his voice through the headset of my APH-5 as he acknowledged clearance to depart. We followed him out.

When we landed at Phu Loi, the line-chief, who had served with Dan and me in Germany, came over to tie down the aircraft. "Mr. Vaughan, did you hear about Mr. Lambdin?"

"What do you mean, hear about him? I just left him. Hear what?"

"He was just shot down and killed near Binh Khat. He was delivering a generator."

"No," I cried out. "That's not possible!"

When I spoke those words I wasn't on the flight line at Phu Loi, I was back at The Wall, but the anguish in my voice bridged the half century that separated now from then.

Nobody paid attention to my cry, they were all dealing with memories of their own.

With my throat choked and my eyes dimmed by tears, I continued my sojourn down The Wall, putting my hands on 22 more names, feeling each of them, seeing them as the young, vibrant men and women I remembered.

Looking around, I realized I wasn't the only one traveling through time and space that day. And for that moment, my fellow time travelers were no longer old men with gray hair and drawn skin. I saw them as young soldiers, and they were wearing jungle fatigues and flak vests with M-16s slung over their shoulders or .45 pistols strapped to their sides or nestled in shoulder holsters. Their eyes weren't dimmed with tears—they were hardened by the thousand-yard stare we wore in-country.

Saying one final goodbye to my fallen comrades, I walked away from the memorial with tears rolling down my cheeks. I left The Wall behind me, but I did not leave the ghosts and the memories. They will be with me until I cross that great divide. Vietnam was but three years out of my 83 years of life . . . but the impact those three years had on me is incalculable.

Transition to the Computer

I started writing on a manual typewriter, and I used nothing but a manual typewriter for the first six years I was writing. Then I got an electric portable and eventually an IBM Selectric. The Selectric was grand, and I thought I was really uptown.

It was 16 more years before I got my first computer, an Apple. There was no hard drive, you had to boot it every time you used it with one of those big, old paper floppies, then take that boot floppy out and put in a different one containing the word processing program. I could get 17 pages on the screen before I had to save it and start a new file. You saved by hitting Control-S, which saved the screen, and you brought something to the screen by hitting Control-L.

But here was the problem. If you hit Control-S when you meant to hit Control-L, you saved a blank screen onto your disk, which destroyed what was on the disc. With the first book I wrote, I would save one chapter and destroy two. It was very frustrating until I finally got the hang of it.

There have been dozens of computers and printers since

then, but every now and then I think back to my days on the typewriter. I got a tremendous sense of reinforcement from seeing the stack of typed pages pile up. I was also closer to the story, because I could take that pile of first-draft pages to a coffee shop, to a table in the back yard, to the Laundromat, even, and edit with a red pencil. That made it more personal, a visceral connection with my characters so that I actually interacted with them.

Don't misunderstand, I would never go back to the typewriter, and in many ways using the computer has made me a better writer because it makes rewriting so much easier.

Ahh, but then there is the internet, a double-edged sword if ever there was one. On the one hand, it is wonderful for looking up information that I might need for the story. If I'm writing about Colorado Springs in 1894, *The Colorado Springs Gazette* for any day in 1894 is available, and I can learn such things as what the weather was for any specific day, whether or not the train was late, what sales were going on in the local stores, and even if there was a fireman's benefit ball.

That is the positive side. The downside is that too often I end up using the internet as a diversion. I will visit a chat room or find newspapers from the 1890s to read—not for research but for fun—or watch cute videos of dogs and other animals (I'm a sucker for them). What had always been my strongest asset as a writer—my work ethic and writing discipline—is being challenged by these distractions. Anything to avoid the often tedious task of writing another thousand or so words.

Miracle on North Kingshighway

What is and what is not a miracle? I've read about things that couldn't be explained: an airplane that flew through the fog, homing in on a signal that wasn't being broadcast, for example. But how many of us have ever experienced anything that we could call a miracle?

I want to share a story that, while possibly not rising to the level of a miracle, is an event I will remember for the rest of my life. It happened a little over 20 years ago, shortly after Pinnacle Publishing went bankrupt. At the time I had a pretty good gig going with Pinnacle, writing romance novels as Patricia Matthews and Paula Fairman.

When Pinnacle went broke, they owed me what amounted to a full year's income. The result was disastrous. I thought I could deduct the money I hadn't been paid as a loss from my income tax, but I learned that if you never received it in the first place, you never actually lost it. As a result, I wound up with a very large—and unpaid—tax bill from the year before.

IRS is very unforgiving. They charge you interest and

penalties, and soon I owed them twice as much as my original debt. They confiscated what money I had remaining in the bank—I didn't know they could do this, but they just took everything.

I was still getting a few royalties from earlier projects, but they were small and far between. I learned not to keep money in the bank but to keep cash so I could buy food and other necessities. I let my other obligations slide, and I started getting nasty telephone calls. It reached the point that when the telephone rang, I would cringe.

After one very difficult call from an angry debt collector, I went over to sit on the couch in my office. I had never been so depressed in my entire life, and I had no idea what to do. I prayed, hard, for something to happen to get us out of this situation.

I looked out the window at the river birch tree, and I gasped. Every leaf on that tree was glowing gold! I don't mean the leaves had turned yellow . . . I mean they were glistening as if sheathed in gold. I know it was the way the sun was hitting them, but none of the leaves in any of the other trees were doing that. It was unbelievably beautiful, and as I sat looking at them—the phenomenon lasted several minutes—a "peace that passeth all understanding" came over me, and the depression lifted. I was a writer, and there were more publishers in the world than Pinnacle. I would just keep writing.

Less than one week later, my agent called. "We have sold *The War Torn* to Dell!" he said. This is a series of five books about WWII, each book from the point of view of one of the belligerent countries: *The Brave and the Lonely* from the American POV, *The Divine and the Damned* from the Japanese, *The Fallen and the Free* from the French, *Masters and Martyrs* from the German, and *The Embattled and the Bold* from the British. The contract was for enough money to get me out of my current condition.

I thanked my agent, and I thanked God.

Was that a miracle? Well, for me, at the time, it was. By the way, for as long as I continued to live in that house, I never again saw those leaves glow in that way.

Oh, by the way, those novels are now available from Wolfpack on Kindle.

Connections Across Time

Writers and readers have a symbiotic relationship that is more pronounced than any other secondary connection. As you read the words, your mind and the mind of the person who wrote them are, for that moment, joined across space and time, whether it is these very words or those of Shakespeare or St. Paul. Writing is only one half the art form. Reading is the very necessary other half.

I have written several historical novels, and I am of the belief that historical novels written with accuracy as to specific events teach history better than a regular textbook. That's because the reader is able to interact, through the fictional characters in the story, with the authentic figures of history. If the story is well told and the reader is fully involved, he or she is traveling through time to relive the events.

The most interesting aspect of writing historical novels is the research. History books are good resources, but so, too, are contemporary accounts. You must take into account, however, that contemporary sources are often filled with inaccuracies, much like today's news stories when journalists, in an effort to make a deadline or be the first to break the story, are forced to go with the information at hand, which may later prove inaccurate. So, to compensate for that, you need several sources for a particular event, and from those sources you can extrapolate the truth.

However, despite the frequent inaccuracies of the newspapers, magazines, and books published at the time, there is something else that is very important, especially to a novelist. When reading an article in a newspaper published in 1870, you are getting the same information that people got at the time. You can understand what they were thinking, what they feared, and what they were hoping for. You have become one with them.

And it isn't just the news stories that are important. Advertisements can teach you can read about the balms, elixirs, and cures, the latest marvels of the times, and much more. And don't pass up the columns. One column in an 1890 newspaper spoke of "Alexander Graham Bell's telephone and Thomas Edison's talking machine. Suppose some clever person could discover how to hook the two machines together so that if one should receive a telephone call while absent, the talking machine could record the caller's message."

Ruth does almost all of my research. She got very interested in the "quaints" written by James Loudin, editor of the Colorado Springs Gazette. Quaints are humorous little stories written in such a way as to mask whether on not they are true. Here is an example of one, titled "The Baptism":

In a pious group there was a Miss Wilson who wanted to be a Baptist, and she presented herself for baptism. Now, Miss Wilson weighed two hundred pounds, including her cork leg, which was a full length leg and modeled in due proportion.

She made an attempt to reach the officiating clergyman breast-deep in the water, but her cork leg was seized with unwanted activity. Miss Wilson knew nothing of the law of specific gravity and was not to blame.

She was suddenly reversed in the water. The minister, feelingly, righted her up and, observing the grinning of the spectators at the solemn scene, asked Miss Wilson to please not do that again.

He was innocently ignorant of the cause of the disturbance of her equilibrium. He gently led the maiden out, when with a shriek, she fell backward, and again her lively leg shot out of the water.

The minister made half-a-dozen efforts but could not keep the convert right-end up long enough to baptize her. At length she told him of her trouble, and he called for a weight to ballast her.

The spectators fled precipitately to give vent to their feelings. Miss Wilson flip-flopped ashore in indignation and amazement and went and joined the Presbyterians.

Every day, Ruth would read another humorous article to me....obviously very much enjoying the research. Then one day I heard her say, "Oh . . . no."

"What is it?" I asked.

"He's dead."

"Who's dead?"

"James Loudin."

"What do you mean he's dead?"

"He died last night," Ruth said, clearly distraught.

"Ruth . . . *he died over one hundred years ago!*"

"But I was reading him yesterday. He died last night."

It took her a couple of days to recover . . . and while on the surface this might sound strange, it is a vivid example of actually putting yourself back in the time and place of whatever you are reading.

Dogs

Some time ago we had to put down our dog, Charley. It was the first time I ever had to do anything like that. My first dog, Suzie, died while I was away in the army. I had a dog, Sheri, when my ex-wife and I separated, but that was over 50 years ago . . . and not only was I gone when Sheri died, I don't even know when it was.

After putting down Charley, I said that I would not get another dog. But I couldn't get over the grief, and ultimately I found another dog that looked exactly like him. We named this one Charley, as well, figuring that if England could have 8 kings named Henry, we could have two dogs named Charley.

Charley 2, like Charley 1, came from a rescue pound. He is a long-hair Jack Russell, and he is very intelligent. You would think that a dog as smart as Charley would understand that we rescued him, and he would be very obsequious and acquiescent out of gratitude. His attitude should be: "I'll do anything you want, because you saved me." He should be humble.

But Charley does not know the meaning of the word humble. He is one of the most arrogant little creatures you will ever encounter. This is *his* house . . . he allows us to live here with him, as his staff. After all, he can't open doors, and he needs us to feed him, give him water, and provide him with toys.

He also needs me to operate the elevator for him. He loves riding in the elevator, and if he hears the doors open from two floors away, he runs up or down the stairs as fast as he can so he can ride the elevator right back to where he was.

When we are watching TV, Charley has his own place on the sofa—and woe betide anyone who tries to sit there. He establishes the schedule and goes out after every meal. It doesn't matter if you took him out just before you ate, he must go out immediately after. He also goes out at three in the afternoon.

I'll be writing, when I feel a paw on my leg. I look down, and Charley is staring at me with occasional glances toward the clock. When I look up, I see that it is three.

Once we are outside Charley will determine where we go. If we go out onto the beach, he will decide whether we are going to walk west or east. And once he makes the decision, I can't change it, unless I want to literally drag his little butt through the sand. If there are people on the beach, he will decide who he wants to go greet and who doesn't interest him. If he decides he wants to speak to someone, he will literally pull me toward them.

Sometimes he doesn't want to go out on the sand but would rather walk up and down the road, carefully examining anything—a rock, a leaf, an insect—that wasn't there the last time he passed by. Frogs fascinate him, and he will put his nose on them, but he won't bother them. The frogs seem to understand, and they don't run from him.

On the other hand, he is death to lizards. He knows all their hideouts and goes immediately to those places to see if any are there. The Geico Insurance gecko is his favorite

TV personality, and I had to explain to him that lizards can't really talk.

Recently scientists have made new revelations about the intelligence of dogs. They have run brain scans of dogs and observed that when a dog sees a familiar person, the brain activity is exactly as it is in a human being. They have concluded that dogs have human-like emotions. They didn't really need an MRI for that; any 10-year-old kid with a dog could have told them.

Do dogs have souls? I am convinced that, yes, they most certainly do. Their actions aren't merely survival instincts. Dogs can love. We've all seen pictures of dogs who actually mourn the death of people or even other animals. Only a creature with a soul can do that.

Do they go to heaven? Well . . . if heaven is supposed to be complete happiness and joy, then I want part of my happiness and joy to be a reunion with Suzie, Sheri, Charley 1, and Charley 2. And I want to be able to talk to them. As Will Rogers once said, "If there are no dogs in heaven, then when I die I want to go where they went."

Friends and Enemies as Characters

I've admitted to not knowing how many books I've written—somewhere between 300 and 400, I suppose. Let's say there are 300, and let's say each one has 30 characters. That means I've had to create about 9,000 people—enough to populate a small town. The hardest part of that is coming up with names. So, I have a confession. If I know you, if I went to school with you, or was in the army with you, or have met you socially or professionally, the odds are very good that you are in one of my books. Sometimes more than one. I also have a habit of using the first name of one friend with the last name of another.

One advantage of being a writer is I can reward my friends—write you in heroic fashion or make you wealthy, handsome, beautiful, athletic. Ahh . . . but for those who may have, over the years, incurred my displeasure, they become the villains, and little do they know that they will spend that very night sleeping in a bed-bug infested hovel. So watch what you say to me, next time we meet.

I Write Westerns Because

"Why do you write Westerns, anyway? Hasn't that become rather an archaic art form? I mean, after all, this isn't the forties or the fifties. Roy Rogers? Gene Autry? Come on, when's the last time you saw kids playing cowboys? Can you even buy a cap gun and holster anymore?"

Every interviewer who has ever asked me that question has done so with a self-satisfied smirk, as if challenging me to defend my genre if I can. Well, I can, though in truth, it needs no defense. It is no surprise that I get mail from readers in England, France, Germany, and even Korea and Japan. Westerns are recognized and appreciated all over the world as the quintessential American art form. What story is more American than a lone wolf with a rifle, a pistol, a pony, and a mission to put things right? There is nothing in literature that is more representative of good versus evil than a cowboy who reluctantly dons the mantle of hero to rescue a damsel in distress, defend a small town from a corrupt sheriff, or save small ranchers from a malevolent land baron.

As I write these stories, I become one of my characters, staring evil down the barrel of a Colt .44. But it is my hope and belief that a retired businessman reading on a beach, an army sergeant in garrison, a truck driver piloting his 18-wheeler along some interstate listening to an audio recording, or a 12-year-old with a love for adventure and a future yet to be realized has also become that Western hero.

A Couple of Missteps

During the run of my American Chronicles series of novels, Bantam arranged a national book-signing and publicity tour. Although this may sound exciting, it is anything but. You have to always consider Murphy's Law. If there is anything that can possibly go wrong . . . it will.

On a radio show on KMOX in St. Louis, I announced that I would be signing books at B. Dalton in Mid Rivers Mall. KMOX has in incredible audience, and when we arrived, there were already dozens of people in line, and the manager and salespeople were scrambling to get things ready.

"We've only got six of your books," the manager said. "We're trying to round up more."

"You've only got six books?"

"We didn't know you were coming today."

"How can you not have known about it? This has been set up for at least two weeks."

"I'm sorry, Mr. Vaughan, we just weren't told."

"It hardly seems worth the trip for just six books," I said, clearly agitated.

As that conversation was going on, Ruth walked to the other end of the mall and discovered a huge sign spread across the front of Waldenbooks: *Meet Author Robert Vaughan Today!* When I walked down to check, I saw a floor dump of my books, a nearby table, pens, a coffee cup, and a comfortable chair.

"Is Robert Vaughan supposed to be here today?" I asked.

"Yes, we're waiting on him now. He should be here by ten."

"Thanks."

Okay, that was my fault: I said the wrong bookstore when I was on the radio program. I hurried back down to B. Dalton, signed six books to the first six people in line, then accepted the apology from the store manager for not being better prepared.

"It isn't your fault. Don't worry about it," I said graciously.

"I must say, you're taking it very well."

"It's all part of the business," I said, showing him how obliging I could be.

By the time I reached Waldenbooks, word had spread through the mall, so most of those who didn't get a book at B. Dalton showed up there.

"It was a mix-up between my publisher and the bookstores," I explained. "It happens. I'm not blaming anyone."

Later on that same tour, I found a beautiful, well-filled basket in the hotel room where I was booked. The basket had chocolate, several kinds of nuts, fruit, crackers, a tin of smoked sardines, and a bottle of wine. I enjoyed it all, and when I was on the phone with my publicist, I asked her to thank Bantam for supplying the wonderful basket.

"Uh, we're paying for the room; I don't know anything about a basket. Let me check." A moment later, she came back on the line. "There is no basket."

"What do you mean? I'm eating from it now."

"You'd better check with the hotel."

I did check with the hotel. The basket wasn't a gratuity . . . it was offered by the hotel—for $50.

At a TV station in Wichita, Ruth and I were waiting in the green room, enjoying donuts and coffee, when a young intern approached.

"Mr. Vaughan?"

"Yes."

"I'm getting the chyron ready, so I need to check on the spelling of your name."

"It's V-a-u-g-h-a-n. Don't forget the last 'a' in Vaughan." I flashed a winning smile. "Vaughans with an 'a' are the aristocratic side of the family."

The joke didn't register as she very carefully spelled my name on a card.

"And why are you here, Mr. Vaughan?"

"Why am I here? I'm going to be interviewed in a few minutes."

"Yes, but what is your title? What do I put under your name?"

"I'm a fandango dancer." I laughed as she walked away. "What a doofus question that was," I said to Ruth. "What does she put under my name."

"That was rather rude of you," Ruth said. "She was just doing her job."

"Look, I've worked in TV. In fact, I used to do this very thing. These shows are all planned in advance. She knows damn well I'm an author. If she doesn't, she should. She'll figure it out soon enough."

A few minutes later, I was on the set being interviewed by some woman who, during the break, asked me what my book was about. I had just over a minute to tell her about it.

The floor director counted down, then pointed, and the red camera light came on.

"My guest today is Robert Vaughan, author of *The Lost Generation*. I thoroughly enjoyed this book. It was a wonderful read. Mr. Vaughan, where did you get the idea for this book?"

"My wife teaches school. I used to give history talks to her class, and that gave me the idea of writing a series of books, each set in a different decade of the twentieth century. This is the third in the series, about the 1920s."

"And will you do books about the other decades of the century?"

"Uh, yes," I replied, realizing she wasn't really paying attention to me.

Whatever the next question was completely escaped me, because at that moment I saw myself on the monitor. Below my picture was the chyron: *Robert Vaughan, Fandango Dancer.*

Okay, I'll admit it . . . just possibly some of these little mishaps might be my fault.

Goats of Oregon

Because of the nature of my business, we've never been tied down to one spot. Once, about 35 years ago, Ruth showed me an ad in a national magazine:

You are invited to live for free for the summer in a rustic, mountain-top cabin, where you will babysit with goats and commune with nature, while the owners visit civilization. An IBM typewriter and a well-stocked personal library will be available for your use. Write an essay as to why you should be chosen.

"You're a writer . . . write an essay," Ruth said.

I did, and two weeks later we learned that we had been selected.

The cabin—and it was a cabin, not a house—had one common room, plus a sleeping loft. It was located in the Blue Mountains of Oregon.

"You are seventy-three miles from gasoline or groceries," a man, who asked that we call him Papa, explained. Papa was a former aeronautical engineer, and his wife, Joanie, was a former English professor at Berkeley. Considerably older than either of us, they had "dropped out" during the sixties and now lived on a subsistence farm, with their only income being the occasional money one or the other made from writing an article.

"I would be interested in knowing how you happened to pick my essay?" I asked, smiling with anticipation. "What did you find most persuasive?"

"You were the only one who responded," Papa said.

After warning that we had to be constantly aware of forest fires, they left us to our own devices. There was electricity and a black-and-white TV set, which got the very weak signal of one channel. They had a pair of old wood-burning stoves: a cookstove and a pot-bellied stove for warmth. The toilet was outside, with no front door but with a beautiful view. The shower, like the toilet, was outside, with nothing to provide privacy but the 73 miles of separation from town or neighbors. The water came from a long hose that snaked down the mountainside from a spring. You had to time your shower just perfectly, to take advantage of the warming effect of the sun on the hose before getting an icy blast from the spring.

And there were two goats—Twileth and Sable—and our only job was to milk them every morning. I tried, but I was totally inadequate to the job. Ruth had grown up on a farm, and she did very well . . . and accused me of purposely failing.

Twileth was pregnant, and when I went out one morning to feed her and Sable, I was greeted by three baby goats . . . not helpless and lying there like newborn puppies, but 100 percent active, bouncing around, jumping up on structures, and ready to play. Our two young boys, Joe and Tom, were with us, and they bonded instantly with the new little creatures. As it turned out, the baby goats enjoyed sucking on their earlobes.

Then I got sick. I'm not talking about a little cold or a slight upset stomach or not feeling well. I'm talking *sick*, as in I was sure I was going to die, sick. I had an insanely high temperature, followed by a chill so severe that it was impossible to get warm. I remember sitting in a chair and noticing that it was about 9:30 in the morning, then glancing up at the clock a minute later to see that it was 7 at night. I had lost an entire day, in the blink of an eye.

We had telephone numbers to call if we spotted a forest fire or if a goat got sick, but nothing for a real doctor. I called a number from a tattered phone book for "transients in need of a doctor."

"What's wrong with you?" the doctor asked.

"I don't know. I'm weak, dizzy, and headachy. Sometimes I'm so hot that I feel like I'm about to catch on fire, then I have a chill where I feel like I'm at the North Pole."

"When you have chill like that, wrap up in a blanket. When you are hot, take it off," the doctor said.

"That's it?"

"What else do you want?"

"Nothing, I guess."

The problem with the illness is it would come for about three days, then go away. Just when I thought it was finally over, it would come back.

When at last it began to ease its grip on me, it struck Ruth. We had no idea what it was, but we were afraid it would hit the kids, so as soon as Papa and Joanie came back from one of their trips, we told them we were leaving and explained why.

Ruth was still going through the same thing I had gone through, so she lay in the back seat of the car, alternately burning up and freezing to death throughout the long drive home. By the time we reached Sikeston, she was in one of those periods where the illness was gone, and it had now been almost two weeks since I had shown any symptoms, so we didn't bother to see a doctor. That night, though, it struck her again, so the next morning she went to see her brother-in-law, who was a pediatrician. He ran tests on her, and then on me, and it turned out that we were both suffering from Rocky Mountain Spotted Fever. We learned, also, that one in four victims die if they get no treatment. We were prescribed massive doses of tetracycline, and I took the prescription to the drugstore.

"Is this for a child? It's quite a large dose," the pharmacist said.

"No, it isn't for a child."

"Still, this is a large amount, even for an adult."

"It's for two adults . . . my wife and me."

"Oh? For both of you?"

"Yes."

The pharmacist smiled. "All right."

A little later, I asked my brother-in-law why the pharmacist smiled.

He laughed. "Well, it was for both of you, and he is probably wondering which one of you gave it to the other. This drug is also what would be prescribed for a sexually transmitted disease."

I noticed that for the rest of the time we lived in Sikeston, that particular druggist never looked at me without a knowing smile . . . and I never disabused him of his notion.

Well, It Could Have Been Him

Forty years ago, the magazine *Scope* that I was publishing for the Sikeston Standard morphed into my own magazine, *Delta Metro*. I really enjoyed working on it. I had a great staff of writers, ad salespeople, and print composition people. *Delta Metro* came as old Linotype was being replaced by off-set printing, and we had machines to set the type and headlines. We also had a dark room and a machine that would photograph our pasteups and produce a full-size page on a negative.

I didn't know the names of any of the machines, nor did any of my people, some of whom had previous newspaper experience. Since we didn't know the names, I made them up. The typesetter became the *dinkle* machine, the headline setter the *double dinkle,* and the big machine that converted pasteup pages to negatives was the *wrinkle-crinch*. Soon everyone referred to the machines that way so we would know what they were talking about.

There was one drawback. A young man who worked for me was going to move to St. Louis and apply for a job at the newspaper there. "Dick, what will I tell them I worked on?" He asked. "They won't know what a wrinkle-crinch machine is." I didn't know the answer, because I didn't know what the machine was, either.

Once the wrinkle-crinch had done its job, I took the negatives over to the *Sikeston Standard*, which had been—forever—the principle newspaper in town. Technically, I suppose, we were competitors, but not really, as the *Standard* was a daily, and they did the up-to-date news. We were a weekly and focused on features more than news. Besides, one of the owners of the newspaper, Allen Blanton, was a classmate and friend of mine.

The *Standard* had the big web press that would print the paper for me. If you are in this business, or if you have ever had anything to do with this business, you will understand that there are few sounds more exciting than the ringing of the bell before the press starts its run. You hear the bell and the heavy rumble of the press, then you watch the paper whiz through, and finally you see the finished product. Even now, as I write this, I can recall those sights and sounds, and I miss that excitement.

One of the features I enjoyed doing was a weekly history story about Sikeston or the Southeast Missouri area. One story was about Thompson's gold. Jeff Thompson was a Confederate general who may or may not have hidden $100,000 in gold somewhere between Sikeston and New Madrid. At today's value, that would be millions of dollars.

I also did a story about a bank robbery that took place in Sikeston in the 1890s. When I set the type, I had some room left over—a spot that would be perfect for a photo, if I could find one suitable for the story. I searched all the archival material I could but was unable to come up with anything. Then I saw a late-1800s picture of the Methodist Men's Bible Study class—three rows of men staring at the camera. In the back row there was man with a stern expression and a beautiful, full mustache. He looked exactly like I imagined a 19th century bank robber would look, so I cropped out the other two men, blew up the photo, and ran it with the story.

"Arthur Jenkins, bank robber," the caption read.

The paper went to press, and soon I was basking in the favorable responses I got from my exciting story about the bank robbery. Then my secretary came to me. "Dick, there's a lady on the phone who wants to talk to you about your bank robbery story."

"Great!" I said, preparing myself for more praise. I'm sufficiently self-centered that I couldn't get too much laudation. I took the call.

"Mr. Vaughan?" the caller asked. I don't mean to be politically incorrect here, but she sounded old—probably as old as I am now.

"Yes, ma'am?" I said, smiling as I awaited her compliments.

"Where'd you get that picture of the bank robber you ran in your story?"

"Why, I got it from some archival photos," I replied.

"Well, that was my father, and I'll have you know that he was a member of the Methodist Men's Bible Study class . . . and he would *never* rob a bank," she snapped, slamming down the phone before I could answer.

It is only now that I will confess where I got that picture.

A Close Encounter of the Frightening Kind

I had a pretty good thing going with Pinnacle, but they went bankrupt owing me a good deal of money and also left me without a book contract.

I was surprised and pleased when a former Pinnacle editor called and asked, "Have you got anything going, Dick?"

"No."

"I'm working for *Playboy* magazine right now. How would you like to write something for me?"

"You mean people actually do buy *Playboy* for the articles?"

"Very funny. There's ten thousand dollars in it for you if you're interested."

"I'm interested!" That was more money than many of my book advances.

"How far are you from the federal prison in Marion, Illinois?" he asked.

"About a hundred miles."

"Then you can get there quite easily to interview a prisoner. I want to do a *Playboy* interview with Edwin Wilson."

Edwin Wilson was a former CIA agent who went rogue,

using his CIA connections to become fabulously wealthy. He owned three airplanes, four homes, a 2,500-acre horse farm just outside Washington, and over 100 businesses. He was also a gunrunner, doing 10 percent of the gunrunning for the CIA and 90 percent for himself.

I called the warden at the prison to see how I would go about getting in touch with Wilson. The warden took my number and said the only way I could talk to him would be for him to call me . . . collect. I agreed to the procedure, and within the hour I received a call from him.

Wilson seemed eager to talk to me and was quite candid in his conversation, even bragging about buying a Ford van for $15,000, putting about $1,500 worth of radios in it, and selling it to Libyan dictator Muammar Khadafi for $150,000. Over the next few weeks he phoned several times, surprising me once by calling me Dick. I had introduced myself to him as Robert and had never told him that I go by Dick.

"What did you call me?" I asked warily.

"I called you Dick. That's what your friends call you, isn't it?"

"Yes. . . ."

"Well, aren't we friends?"

I could almost hear him grinning, and I let the matter drop.

It finally reached the point where I felt I needed to see him in person. Also, *Playboy* would want some photos, so I called the warden and made an appointment to see him. But when I arrived, the warden denied my visit.

"Why can't I see him?"

"Mr. Wilson has been a bad boy."

"Warden, I may be wrong, but I thought everyone in the Marion maximum-security federal prison is a bad boy."

As it turned out, I wasn't the only journalist Wilson was speaking with. Without arranging it with the warden, he had invited a BBC television crew to interview him. They flew over from London and showed up at the gates of the prison, only to be turned away.

I continued my conversations with Wilson, because he continued to call, each time getting more *friendly*—and to be honest, just a little more frightening.

"Ed, have you ever killed anyone?" I asked during one of the calls.

"Well, not in America," he replied.

Then one day he said, "I see where your son Tom got most valuable player in the basketball tournament in St. Louis. You must be proud of him."

That brought me to a complete halt. I had never mentioned my family to him. Also, I had adopted Ruth's two boys after we were married, but because they were in school, they opted to keep their birth name. How did Wilson connect Tom with me?

Except for my telephone number, I had given him no information about where I lived, yet during another call he said, "So, you're in Sikeston, Missouri, on North Kingshighway, aren't you? You live in that two-story house, right across from the library?"

The whole thing was getting a little too weird and a little too close, so the next time he called, I figured out a way to withdrew from our discussions.

"Ed, I'm not going to do this article, and here's why. *Playboy* wants me to do a hit piece on you . . . and I just won't do it. I feel like we have become friends over the last several weeks. I know you have an appeal in, and I don't want to do anything that would hurt it."

Wilson was silent for a moment, and I waited with some apprehension to see how he was going to react.

"Dick," he said, "I . . . I don't know how to thank you for that. You are one of the most decent people I've ever encountered. I know you must be giving up a lot of money to back away . . . and if I had the resources I once had, I would more than make it up to you. Thank you. From the bottom of my heart, thank you."

I almost felt bad about lying to him . . . though it wasn't

a complete lie. I had been in conversation with *Playboy*, and they did want an *edgy* piece.

Many years later, Ed was released from prison when a federal judge granted his appeal. He has since died, and though his lawyer admitted that Ed was guilty as charged, his release was granted because the CIA manufactured evidence used to convict him.

Of Riper Years

I have heard it said that it sucks to be old. Well, that may or may not be true, but once you are old, there is nothing you can do about it. And, when I think of how many of my friends and family never had the chance to get old... then it is a condition I am not only thankful for, it is one that I relish.

Those of us who are of "riper years" have built a treasure trove of memories that are unique to us—memories that, for some strange reason, seem to get sharper as we age. And we can, anytime we wish, pull those memories out and relive them... not as some life-changing event, but as individual snippets of recollection. They are, in fact, more enjoyable when contemplated as totally unrelated events.

I'll share some with you:

A brand-new 1949 Chrysler parked too close to a Little League baseball game, having one of its headlights broken by a foul ball.

Seeing the fireman or the engineer use a huge oilcan to oil the driver wheels of a 4-6-4 steam locomotive as the train stood in the station, taking on passengers, at the Frisco Depot in Sikeston.

Looking at the downtown Christmas decorations and lights through the back window of the car, after a rain, each of the clinging droplets gleaming with the colors of the lights.

The smell of oiled "floor sweep" on the floors of South Grade School.

The first time I ever went up in an airplane . . . with my dad flying.

Eating fried fish sandwiches at the Cotton Carnival.

The feel of cotton in your fingers as you pull it from the boll, dragging your sack behind you.

Crossing the Mississippi River on the old Frisco Bridge in Memphis.

The sound of a football being kicked in the pregame warm-ups, and the sight of the goalposts being decorated with crepe paper in the colors of the schools that were playing.

Seeing the ocean for the first time.

My dog Suzie.

My parrot Bud.

Playing pitch and catch with my brother Tommy, who has been gone for many years now.

Playing some game that my best friend and I made up.

Being on the KD (known distance) firing range in the army, feeling the recoil of the M1 rifle, then looking toward the target, waiting for the target pullers to mark your target with the marker disk on the end of a long pole, praying that you don't get a Maggie's Drawers and teasing your friends when they do.

Being on KP and sitting in the mess hall at 0400 during those few minutes before all the work begins, dozing and listening to the gurgling sound of the coffee percolators.

Holding my first son, Ricky, for the first time.

Selling my first book.

Flying solo for the first time.

The complete and total joy of having a warrant officer's bar pinned to my uniform.

Seeing a string of green lights coming up from the ground and realizing, with a start, that they are tracer rounds being shot at my helicopter. The joy of having them miss.

Returning home after an overseas deployment and seeing the excitement of my dog, Sheri, who hadn't forgotten me.

The sound of the warning bell when the web press started to print the first edition of my news magazine, *Delta Metro*.

Watching my son run in an AAU track meet when he was 9 . . . then years later, watching him running track for Washington U.

Sitting in a noisy gym in Oregon watching my youngest play basketball as a fifth grader . . . then years later, watching him play for Westminster College.

And finally, the awful memory of taking Charley 1 to the vet on that terrible, last visit . . . holding him one last time. And yes, painful though that memory is, it is still precious to me.

Ancient Greek and Drinking Coffee

I never drank coffee growing up. That meant that I never drank coffee throughout my entire army career, and I used to envy those who did. There was always coffee available somewhere. When we were in the field in near zero weather, there would be a marmite can of hot coffee, and the men would line up for it. They obviously enjoyed it if for no other reason than the temporary warmth it provided. In the meantime, I would shiver in the snow.

It may be that the reason I didn't drink coffee growing up was because of the way my mother made it. She was from Jackson, Mississippi, and made coffee so strong that all of her friends teased her about it. "Aline, you don't want to keep a spoon in the cup too long, the coffee would eat it."

My mother's sister, Aunt Eunice, came to stay with us for a while, and she brought a little jar of instant coffee. I thought, *Hmm, she knows how strong my mother's coffee is.* I was wrong. She dropped a spoon of instant coffee into the cup of Mother's coffee and stirred, saying, "Aline, since you've come north, you've forgotten how to make coffee.

This isn't anything more than brown water." I hasten to add here that *north* was Sikeston, in the Bootheel of Missouri, and we didn't really consider ourselves Yankees.

I do drink coffee now. I like it very much and regret all the years I did without. I didn't start until I was in my seventies and had moved down to Gulf Shores, Alabama. I would walk Charley on the beach every morning, then a few of us would gather for coffee at the house of one of our neighbors. I forced myself to drink it, because I didn't want to be the odd man out.

It was an interesting group of men who sat around and told stories over our cups of coffee. One was captain of a charter fishing boat. One had once worked on the offshore oil rigs. One was an occasional coffee drinker named George who actually lived in Mobile but had a house on the beach that he visited from time to time.

"How's your latest book coming along, Dick?" George had to remove a toothpick from his mouth to ask the question.

I knew he was just being nice to me and didn't figure him for much of a reader. Oh, he might have read some of my Westerns, but I was working on a tome that was considerably more literary than my normal books. I was writing *The Masada Scroll,* a book in the same genre as *Quo Vadis* or *Ben Hur.*

"Fine," I said. I decided to impress him with the depth of my book. "I'm stuck for the moment, though. I can find a translation for Greek, but I can't find one for ancient Greek, and I really need to use ancient Greek for a scene."

George pulled the toothpick from his mouth, drawing it out like some farmer leaning against the back wheel of a tractor he had just climbed down from after planting a field of soybeans. "Well, I can translate it for you."

"What?" I asked with a laugh, absolutely certain he was putting me on because of my *I'm a writer* attitude. "You can translate ancient Greek?"

"Ancient Greek is one of the courses I teach at Springhill College." He put the toothpick back in his mouth and smiled.

Professor George was as good as his word. When the book was published, the scroll of *The Masada Scroll* was translated into ancient Greek.

Football

Recently I had a doctor's appointment, and perhaps because I am as old as I am, I had to fill out a wellness survey in which I was asked a lot of questions, some of them quite personal. One was, *Are you depressed?* I answered, "It depends on whether Alabama wins or loses."

Well, Alabama lost to LSU, and the game turned on two plays: Tua Tagovailoa's fumble inside LSU's 10-yard line, and an intercepted pass that was called back because of an illegal-participation penalty against Alabama. Both plays resulted in LSU scores.

Now, don't get me wrong, I am not blaming Tua for the fumble. That's part of football, which I know because I have fumbled a football or two. I'm also not disputing the referee's call of illegal participation. But the bottom line is, Alabama lost, and that depresses me.

I think I need to give you a little personal history. Why is someone who was born and raised in Missouri an Alabama fan? It started in 1947, when I was 10 years old. We didn't have TV then, and the games broadcast over the radio were hit and miss as to which ones you could hear. But we could follow our favorite teams in the sports page of the Sunday newspapers.

I was visiting Tom Murchison, my best friend then and someone who remained my best friend for an entire lifetime. "Ah, Tennessee won yesterday!" he said triumphantly, reading the results. "Tennessee is my favorite team."

"They're my favorite team, too," I said. I must point out that Tom—Tommy back then—was not only my best friend, he was smarter, a better athlete, and more popular than I was, and I looked up to him. I chose Tennessee because I thought it would please him.

"No, Tennessee is my team. You have to get your own team."

I looked in the paper and saw that Alabama had beaten Georgia. And I had lived in Alabama, even attended school there during the war, when my dad was at Fort (then Camp) Rucker.

"Okay," I said. "My favorite team is Alabama."

That random pick has given me over 70 years of satisfaction, because Alabama has had some great teams. They've won the national championship 17 times.

But I'm 83 years old. Why do I get depressed over something that a bunch of 18- to 20-year-old kids do? Well, depression over losing a football game has long been a part of my psyche. In 1953, Poplar Bluff beat Sikeston, 14 to 13. That was the first game Sikeston had lost since 1947.

And I'm still depressed over that game.

Honored to Do This One

Colonel Ben Purcell, who was the highest-ranking army officer to be a POW of the North Vietnamese, recently died. I had ghost written his book *Love and Duty*. When I was first contacted to do the book, I traveled to Georgia to meet him. He was scheduled to speak at a Baptist church, so I went to listen.

There were several teenagers in the congregation, and listening to someone speak was not very high on their agenda. As Ben started, they were still fidgeting and teasing one another. Ben wasn't a very dynamic speaker—he spoke quietly and with little fluctuation in his delivery—but his story was immensely powerful. Within two minutes, the young people were as quiet and still as the rest of the congregation. And by the time Ben finished, there wasn't a dry eye in the house . . . including the young people. I knew that night I wanted to write Ben's story.

Ben was skilled with his hands. As a POW, he manufactured a key by cutting several strips in a flattened toothpaste tube, inserting and turning it in the lock, then withdrawing it to see which of the little strips had been pushed out by the tumblers. From that, he manufactured a key that allowed him to escape his cell and the prison, only to be betrayed by the French Embassy where he had gone, thinking they could get him out of the country.

Here's an example of how detail-oriented he was. I arrived at his house one day to do some work on the book.

"How was your trip?" Ben asked.

"It was fine, but something happened to my car door, and I can't lock it, not with the remote or manually."

Ruth and I went into the house and visited a while with his wife, Ann, who had a big part in the book. *Love and Duty* book is written in alternating chapters: what Ben was doing in prison and Ann was doing at home.

"Where's Ben?" I asked.

"Oh, he's out in the driveway, working on your car."

Working on my car?

I went outside and was shocked to see that my door was off the car, separated into two pieces, with scores of very little parts spread out on the driveway. Ben was in the garage, standing be the vise and using a file.

"Ben?"

"Here's your problem," he said, pointing to a piece lying on the table. He went back to his filing. "I have to make you a new one."

By noon the door was reassembled, back on the car, with the lock working perfectly by remote and manual.

We finished the book, and I submitted it to potential publishers. One of them, St. Martin's Press, invited us to appear before their editorial board, which is very rare. I asked Ben to tell the handshake story, which I'm going to share here.

Ben wasn't kept in the Hanoi Hilton with the other American prisoners. He was the only American in his prison, because the NVN already had an air force colonel named Ben Purcell, and they thought Ben must be a CIA agent using the same name. Because of that, he was kept in solitary confinement except when he was being interrogated ... so he would go for months without encountering another human being.

One morning, Ben cleaned his cell, sweeping the debris under the door, which had about a 10-inch gap between the bottom and the floor. The prison had Vietnamese prisoners, and that morning one of them was sweeping out in the hallway, and he swept the debris back into Ben's cell. Ben swept it out again, and the Vietnamese prisoner swept it back in. Ben swept it out a third time, and then he got down on his hands and knees, and when the prisoner started to sweep it back in, Ben grabbed the broom and jerked it away.

The prisoner stuck his hand under the door, asking for the broom to be returned. For a moment, Ben considered breaking the broom. Then he thought that this guy was a prisoner, just as he was, so Ben returned the broom. The prisoner put his hand back under the door, hit the floor once, then offered it as a handshake. As Ben took his hand, he realized that it was the first time he had touched another human being in over a year. He held onto the prisoner's hand ... and the prisoner, realizing this, reached under the door with his other hand and stroked Ben's.

Ben could not tell that story without tearing up. And believe me, I couldn't listen to it without tearing up, nor could anyone on that editorial board. St. Martin's bought and published Ben's book.

Partying with Perot

During the Vietnam War, Ross Perot spent a lot of time in France, trying to get the POWs released. Because of that, he had a party for the former prisoners at his Texas home.

To keep too many cars from showing up, we had rendezvous positions all over town. Ben and Ann, along with Ruth and I as their guests, met in a hotel where we would board a bus to take us out to the ranch at five o'clock, which of course is rush hour. But when the bus turned onto the LBJ, which is probably Dallas's most traveled route, the highway was empty! It had been cleared for us, because we were going to Ross Perot's POW party.

Ann Purcell had gone to France with Perot on one of his trips, so Ross knew Ann and greeted her warmly. Ben introduced me as the person who wrote his book, and Ross shook my hand and thanked me for my service.

An hour later, when the party was in full swing and everyone was eating grilled steaks, Ross came to our table. "How's it going, Dick?" he asked.

"Everything's fine. I've been able to talk to a lot of former

POWs. I'm awed by it."

When Ross Perot left the table, Ruth turned to me with a surprised look on her face. "He remembered your name."

"Big deal. I remembered his."

A Moving Moment

I was in Costco, sitting at one of the tables in the refreshment area. I was wearing my Vietnam Veterans' cap when a very attractive young woman who couldn't have much over 20 came to my table. She had a baby in her arms.

"Sir," she said. "I want to thank you for your service in Vietnam."

I was moved by that, considering that the final thunder of artillery, staccato sound of machine-gun fire, and *whop-whop-whop* of Huey rotor blades happened long before she was born.

This is a paean dedicated to all who, throughout the history of our country, have worn the uniform of service, and particularly those who gave their last full measure of devotion.

I am 83 years old, and the shadows of my life have lengthened as the time I have remaining grows short. It has been half a lifetime since I last heard the sound of the bugles blowing *Reveille, First Call, Chow Call,* or the haunting sound of *Taps* that put us to sleep at night. I have also blown

Taps to mark the permanent end of duty for those men and women who have gone on to their final roll call.

My time in the army, though but one-quarter of my life, remains as immediate to me as KP at Fort Rucker, duty officer at Fort Riley, and the monsoons of Vietnam. I remember the men in the barracks with me at Fort Leonard Wood, those who climbed the mountains with me on those frigid field exercises in Korea, the morning runs at Fort Campbell, the camaraderie of the officers' club in Germany, my students in the Aviation Maintenance Officers Course at Fort Eustis, and those men who served with me in Vietnam, some of whom have their names inscribed on that shining black wall in Washington and forever in my heart.

A history of military service for men of my age isn't unique, as the draft was a part of our young lives. And though many served their two years, then out, whether they clerked at Fort Benning, worked the flight line at Scott Air Force Base, or did sea or shore duty at Norfolk, all have my respect. And I hold in esteem and admiration all my fellow vets, those remaining from WWII, Korea, the Cold War, Vietnam, and this long, long war against terror. You are all loved.

Milestones

As a writer, I have purposely sought as many experiences as I could, believing they are the stock in trade to a writer, just as groceries on the shelf are stock in trade to the grocer. Some of you may remember the *Time Magazine* feature *Milestones*. I regard these experiences as milestones in my own life.

I will start when I was five years old. I lived across the street from Gerald and Ray Kelly, and they are earliest friends that I can remember. Gerald is gone now, but I'm happy to say that Ray is still around. I also lived across the street from Mr. Clark, who was a Civil War veteran. I have to admit that the fact he was a Civil War veteran meant nothing to me then. I knew him only as a very old man who used a cane to walk to the corner and back. I also confess that I don't know which side he was on, as this was in Missouri, and Missouri literally was a state where brother fought against brother.

I remember WWII vividly. My dad was drafted, and we followed him to Fort Rucker (then Camp Rucker) in Alabama.

At Camp Rucker we rented a bedroom and shared the kitchen with the family that owned the house. The house had no telephone, no electricity, and no water, except for the well. It had no radio, of course, so I remember my mother driving the car up close to the porch, then turning on the car radio so we could hear the latest news. The school bus stopped in front of the house and took me to school in Ozark.

From Fort Rucker we went to Fort Sill, Oklahoma. There, I went to school in Lawton, and at a presentation I gave recently, one of the attendees and I started talking. His dad was also at Fort Sill during the war. He, too, went to school in Lawton and by coincidence had the same teacher there. An even bigger coincidence is that we both graduated in 1955, which meant we were classmates in the same room!

When Dad first deployed, we lived for a little in my mother's hometown of Jackson, Mississippi, where I went to school at Galloway. My teacher, Mrs. Broome, had also been my mother's teacher.

It was there that I gave my first public speech. We had an Easter event attended by all the parents and family, and I was the narrator. Mrs. Broome kept telling me not to be afraid, but I had no idea what she was talking about. Why should I be afraid?

After that school year ended, we returned to Sikeston. I remember the blackouts and being absolutely certain that Japanese airplanes would come from the west and German ones from the east to rendezvous over our small town, because in my mind, at least, we had to be a prime target.

Then came the end of the war. Mother took my two brothers and me downtown to see the celebration. There was no official celebration, but people were shouting, *"The war is over!"* Car horns were honking, and one man, standing out on a balcony, was firing a shotgun into the air.

I remember the excitement of the new cars coming out after the end of the war, and I was very disappointed that we didn't get one—a Plymouth—until 1949. In 1950, Little League baseball came to Sikeston, and I played catcher for the Dons. I didn't like the team name, which meant nothing to me—though I later learned it meant that, unlike the sponsored teams, our team was the result of donations. The good part about playing for the Dons was that our coach was Bill Puckett, who was "Mr. Baseball" in Sikeston.

My dad owned two large trucks, and the summer after I graduated from high school, I became an assistant driver, making runs from Miami to Milwaukee. It was exciting to be driving a big 18 wheeler while the other driver was asleep in the sleeper cab behind me.

I attended one very lackluster semester of college at Rolla, but "honorably withdrew" from every class except English, PE, and ROTC at the end of the first quarter. And though I didn't fail any classes, I had credit only for those three at the semester break.

I joined the army and participated in three significant historical events.

While in Korea, we did a landing of the type carried out on D-Day. It was a practice landing, but we went over the side, crawling down cargo nets into the landing barges and hitting the beach just as if it were the real thing. We weren't under fire, of course, but that certainly provided a perspective for the D-Day scene I wrote in *Portals of Hell*.

My second historical event was the Cuban Missile Crisis. Along with 20,000 other troops, I waited at Homestead Air Force Base for orders to invade Cuba. Fortunately, those orders never came, but I well remember the tension, and I was able to call upon that feeling in *The New Frontier*.

The third historical event was my time in Vietnam. I did three tours and experienced being shot at, being mortared, being damn scared, losing close friends, and the being exhilarated at returning safely to the States after completing a tour. I also experienced the shock of the hostility shown returning Vietnam vets by Americans at home.

My Vietnam experience led to several books: *Brandywine's War*, *Brandywine's War: Back in Country*, *The Valkyrie Mandate*, *The Quick and the Dead*, *Flower Children*, *Dateline: Phu Loi*, and *Dateline: An Loi*.

As I have reached the twilight of my years, I no longer seem to be collecting new experiences, but the real value of an experience is that it is with you, always. I have but to close my eyes, and I can see majestic beauty of the Rocky Mountains, the rooftops of London, the castles along the Main River in Germany, friends on the flight lines at Fort Campbell, in Korea, and Vietnam.

I have shaken hands with three presidents, shared a limo with Pearl Buck and a drink with Norman Mailer, took a class from Joseph Heller, and taught one to John Grisham. I have one surviving brother, four adult children, four granddaughters, one grandson, and two great-granddaughters. I have an aunt, cousins that I grew up with, friends I have known for well over 70 years, friends that I have known but for a year, and distant internet friends that I have never met in person. I've known a number of women, had relationships good and bad, and they all led me to the greatest one of all—45 years with my best friend and love, Ruth. And even as I type these words, I have a dog curled at my feet.

Life is sweet.

About the Author

Robert Vaughan sold his first book when he was 19. That was 57 years and nearly 500 books ago. His books have hit the NYT bestseller list seven times. He has won the Spur Award, the PORGIE Award (Best Paperback Original), the Western Fictioneers Lifetime Achievement Award, received the Readwest President's Award for Excellence in Western Fiction, is a member of the American Writers Hall of Fame and is a Pulitzer Prize nominee.

Vaughn is also a retired army officer, helicopter pilot with three tours in Vietnam. And received the Distinguished Flying Cross, the Purple Heart, The Bronze Star with three oak leaf clusters, the Air Medal for valor with 35 oak leaf clusters, the Army Commendation Medal, the Meritorious Service Medal, and the Vietnamese Cross of Gallantry.

www.ingramcontent.com/pod-product-compliance
Lightning Source LLC
Chambersburg PA
CBHW011343090426
42743CB00019B/3424